Extractive Metallurgy of Rare Earths

Extractive Metallurgy of Rare Earths

Editor

Gyan Singh

scitus academics

Extractive Metallurgy of Rare Earths
Edited by **Gyan Singh**

ISBN: 978-1-68117-678-9
Library of Congress Control Number: 2016958832

SCITUS Academics LLC,
616, Corporate Way, Suite 2, 4766,
Valley Cottage, NY 10989

www.scitusacademics.com

Notice

Printed in United States of America on Acid Free Paper

Preface

We live in a material world. Today, it is the role of the materials engineer to study, develop, design and operate processes that transform raw materials into useful engineering products intended to improve the quality of our lives. The industrial revolution thrust metals into the forefront of technology, and they have become the very foundation of our modern society. One cannot envision a life in which electronics, transportation systems, buildings and machines are not part of our daily lives. Metallurgy is the part of materials science and engineering that studies the physical and chemical behavior of metallic elements, their intermetallic compounds and their alloys.

Rare earths are a series of chemical elements found in the Earth's crust that are vital to many modern technologies, including consumer electronics, computers and networks, communications, clean energy, advanced transportation, health care, environmental mitigation, national defense, and many others. Because of their unique magnetic, luminescent, and electrochemical properties, these elements help make many technologies perform with reduced weight, reduced emissions, and energy consumption; or give them greater efficiency, performance, miniaturization, speed, durability, and thermal stability. Rare earth-enabled products and technologies help fuel global economic growth, maintain high standards of living, and even save lives.

Extractive Metallurgy of Rare Earths accumulates information from disseminated sources that is frequently available only to specialists. It delivers a complete and usable survey of the rare earth resources, extraction, and production of numerous end products. The book deals with the cutting-edge developments in rare earth extraction, production, and processing technology, design, and processes in the preparation and purification of rare earths from minerals and ores to useful compounds and alloys.

In technologies including smartphones, flat-panel TVs, hybrid cars and wind turbines, a number of elements with exotic names represent crucial components. But the elements' rapid rise from lab curiosities to helping supply the hearts of modern electronics has put strain upon existing supply chains, and experts worry that any supply disruptions could slow the pace of innovation. Now countries all over the world have begun racing to secure new reserves, boost recycling and find material substitutes for these energy-critical materials.

Table of Contents

Chapter 1	Rare Earth Elements: Overview of Mining, Mineralogy, Uses, Sustainability and Environmental Impact	1
Chapter 2	Geological, Mineralogical and Geochemical Aspects for Critical and Rare Metals in Greece	35
Chapter 3	Global Production Estimation of Rare Earth Elements and Their Environmental Impacts on Soils	61
Chapter 4	Analysis of Different Rare Metals, Rare Earth Elements, and other Common Metals in Groundwater of South West Bank/Palestine by ICP/MS-Data and Health Aspects	77
Chapter 5	Modeling Preparative Chromatographic Separation of Heavy Rare Earth Elements and Optimization of Thulium Purification	95
Chapter 6	Characterization and Iron Removal Treatment of Ion-Adsorption Rare Earth Tailings in Southern China	113

Chapter 7 Normalized Rare Earth Elements in Water, Sediments, and Wine: Identifying Sources and Environmental Redox Conditions 127

Chapter 8 Separation of Kaolinite from Ion-Adsorption Rare Earth Tailings in Southern China and Iron Removal Treatment 163

Chapter 9 Dissolution of Rare Earth Elements from Coal Fly Ash Particles in a Dilute H2SO4 Solvent 177

Chapter 10 Rare Metal Deposits of East Kazakhstan: Geologic Position and Prognostic Criteria 195

Chapter 11 Optimization of Acid Leaching of Rare-Earth Elements from Mongolian Apatite-Based Ore 211

Chapter 12 Separation and Recycling for Rare Earth Elements by Homogeneous Liquid-Liquid Extraction (HoLLE) Using a pH-Responsive Fluorine-Based Surfactant 239

Chapter 13 Use of Phosphate Solubilizing Bacteria to Leach Rare Earth Elements from Monazite-Bearing Ore 255

Chapter 14 Porosity and Permeability of Round Top Mountain Rhyolite (Texas, USA) Favor Coarse Crush Size for Rare Earth Element Heap Leach 277

Index 305

1

CHAPTER

Rare Earth Elements: Overview of Mining, Mineralogy, Uses, Sustainability and Environmental Impact

Nawshad Haque [1], Anthony Hughes [2], Seng Lim [3] and Chris Vernon [1]

[1]*Mineral Resources Flagship, CSIRO, Bag 312 Clayton South VIC 3168, Australia*

[2]*Manufacturing Flagship, CSIRO, Bag 312 Clayton South VIC 3168, Australia*

[3]*Energy Flagship, CSIRO, Bag 312 Clayton South VIC 3168, Australia*

ABSTRACT

Rare earths are used in the renewable energy technologies such as wind turbines, batteries, catalysts and electric cars. Current mining, processing and sustainability aspects have been described in this paper. Rare earth availability is undergoing a temporary decline due mainly to quotas being imposed by the Chinese government on export and action taken against illegal mining operations. The reduction in availability coupled with increasing demand has led to increased prices for rare earths. Although the prices have come down recently, this situation is likely to be volatile until material becomes available from new sources or formerly closed mines are reopened. Although the number of identified deposits in the world is close to a thousand, there are only a handful of actual operating mines. Prominent currently operating mines are Bayan Obo in China, Mountain Pass in the US and recently opened Mount Weld in Australia. The major contributor to the total greenhouse gas (GHG) footprint of rare earth processing is hydrochloric acid (*ca.* 38%), followed by steam use (32%) and electricity (12%). Life cycle based water and energy consumption is significantly higher compared with other metals.

Keywords: rare earth; mining; mineralogy; recycling; life cycle assessment

1. INTRODUCTION

Rare earth elements (REEs) include the lanthanide series of the periodic table from atomic number of 57 to 71 starting with lanthanum (La) to lutetium (Lu) and including scandium (Sc) and yttrium (Y). They are in short supply internationally with domination by China in production and trade. The criticality level of REEs has been ranked as high with the highest score of 29 based on the analysis of several reports and three criticality factors [1]. This score reported in this document was calculated by summing the individual scores for each commodity in each of recent studies of materials criticality in the UK, EU, US, South Korea and Japan.

One of the members of the lanthanide series, promethium (Pm) is radioactive and exceedingly scarce. Pm is usually sourced through nuclear transformations. Rare earths are further divided into the light rare earth elements (LREE) and heavy rare earth elements (HREE) with the divide falling between the unpaired and paired electrons in the 4f shell [2]. This divide is somewhat arbitrary and the common convention in industry considers that the LREE includes from lanthanum to europium and includes scandium. The HREEs include from gadolinium to lutetium and includes yttrium. The International Union of Pure and Applied Chemistry (IUPAC) definition includes gadolinium within LREE and based on that all of the HREE have paired electrons. There is some inconsistency in reporting REEs. Companies report lanthanum to neodymium as

LREE and sometime include up to samarium but other times HREEs start from samarium. Sometimes precise classification of rare earth oxides are grouped as LREE La to Nd or Ce, Sm to Gd as medium REEs and Tb to Lu and Y as HREEs or yttric [3].

Over the last 40 to 50 years there has been extensive speculation on peak oil, *i.e.*, the point at which oil production will start to decline. This "peak" concept is now discussed with respect to many elements in the periodic table where the demand is increasing rapidly but the supply is limited [4,5]. Phosphorus (P) and indium (In) fall into this category. In the case of phosphorus, intensive farming practices are required to feed a growing world population but limited phosphorus supply suggests a looming crisis [5]. Indium is a key ingredient for displays in mobile phones and newer generation computers. While it is relatively abundant in the earth's crust, there are few rich mineral deposits meaning that the lifetime of known, minable reserves may be as short as 10 years. The increasing demand for rare earth elements and limited reserves for some rare earths cause similar speculation for future availability. The CSIRO Minerals Down Under Flagship Cluster research used a model called "Geological Resources Supply Demand Model (GeRs-DeMo)" for several minerals (e.g., copper, gold, lithium) that predicts supply and demand of a selected commodity. A peak supply year can be estimated for a mineral commodity using the concept of "peak oil" [6,7]. This model may be applied in the future to predict the sustainability of mixed REEs rather than individual 17 elements that may be complex.

It is likely that wind energy and electric vehicles will be considered as part of the solution for a more sustainable future. Present technologies for electric vehicles and wind turbines rely heavily on dysprosium (Dy) and neodymium (Nd) for rare-earth magnets. Future large scale adoption of these technologies will increase the demand for these two elements [8]. It is anticipated that recycling and recovery will assist to satisfy the demand for rare earth element but it is currently very low (*i.e.*, less than 1%) and will pose significant challenge in terms of collection, processing and their recovery due low concentration in the products [9]. Environmental impact indicators relevant for rare earths were listed and some literature reported results were published [10]. However, the reporting of comprehensive life cycle based assessment results of all relevant environmental impacts for specific rare earth product still non-existence.

There is limited comprehensive information in the open literature on the current status of REEs, partially because during the rise of industries that relies on REEs, China was the major source and the remainder of the world has not until recently actively sought to discover new reserves. The recycling aspects have also been described. This paper is considered as a review since a lot of information have been collected, collated, compiled and analysed. However, a small case-study on environmental impact has been used for demonstration rather than reporting the magnitude of results with certainty to indicate that this aspect is important consideration for rare earth industry.

The objective of this paper is to describe a review on the overview of REE mining, mineralogy, extraction processes, and selected environmental impacts.

2. LITERATURE REVIEW AND ENVIRONMENTAL IMPACT

This paper is based on literature review in the area of rare earth minerals. The literature in various databases and websites have been compiled and analysed critically. The key literatures were identified and data were compiled from these openly available materials.

The expected environmental impact in relation to rare earth processing particularly associated with waste management and disposal of tailings will require change in processing to retrieve rare earths [11]. Recently, a review was published on sustainability aspects of rare earth production [12]. They have provided some guidelines but did not report any specific results for rare earth production processes. There are limited studies on the reported impact of rare earth processing with any great detail. One preliminary study was undertaken with a limited scope [3].

3. RESULTS AND DISCUSSION

3.1. Abundance

Typical abundances for rare earths in the earth's crust are provided in Table 1. Ce, at 60–70 ppm, has a similar abundance to Cu. Of course, the actual availability of these rare earths relates to their mineralogy, since having rich ore bodies means that these elements can be mined and processed more easily than if they are distributed evenly within the earth's crust. Mining of poorer quality ore bodies leads to processing of large amounts of ore, increased energy consumption and CO_2emissions. Two papers discussed this in the context of gold mining [13,14].

Table 1. Abundance, resources, production and uses of the rare earths.

Element	Crustal Abundance in ppm	Resources tonnes	Production tonnes/annum (Year's of Reserve)	Uses, Source of Data and References
Lanthanum (La)	32	22,600,000	12,500 (1,800)	Hybrid engines, metal alloys [15], catalysis, phosphors [16], Carbon Arc Lamps, cigarette lighter flints [17]
Cerium (Ce)	68	31,700,000	24,000 (1,300)	Catalysis particularly auto [18] petroleum refining, metal alloys [15], Phosphors [16], corrosion protection, Carbon Arc Lamps, cigarette lighter flints [17,19]
Praseodymium (Pr)	9.5	4,800,000	2,400 (2,000)	Magnets [15], Optical Fibres, Carbon-Arc Lamps [17]
Neodymium (Nd)	38	16.700,000	7,300 (2,300)	Catalysis particularly petroleum refining [15], hard drives in laptops, headphones, hybrid engines, Nd-Fe-B magnets [17]
Promethium	NA	NA	NA	Nuclear Battery, Pm [17]
Samarium (Sm)	7.9	2,900,000	700 (4,100)	Sm-Co magnets , IR absorption in glass [17]
Europium (Eu)	2.1	244,333	400 (610)	Red colour for TV and computer screens [15], (5%Eu, 95%Y), green phosphor (2% Eu) [16]
Gadolinium (Gd)	7.7	3,622,143	400 (9,100)	Magnets, nuclear magnetic resonance imaging, phosphors [17]

Terbium (Tb)	1.1	566,104	10 (57,000)	Phosphors [16] particularly for fluorescent lamps [17], magnets [15].
Dysprosium (Dy)	6	2,980,000	100 (29,800)	Magnets [15], hybrid engines [15,17], can increase the coercivity Nd-Fe-B magnets.
Holmium (Hm)	1.4	NA	10	Glass colouring agent, lasers [15]
Erbium (Er)	3.8	1,850,000	500 (3,700)	Red, green phosphors, amplifiers for optical fibres transmission, pink in glass melts, sunglasses [17].
Thulium (Th)	0.48	334,255	50 (6,700)	Medial X-ray units—X-ray sensitive phosphors [17]
Ytterbium (Yt)	3.3	1,900,000	50 (38,000)	Lasers, steel alloys—grain refiner [15,17], stress sensors [17]
Lutetium (Lu)		395,000	NA	Catalysts in petroleum refining, Ce-doped Lu-glass used in detectors for positron emission tomography—PET
Yttrium (Y)	30	9,000,000	8,900 (1,011)	Red phosphors [16], fluorescent lamps, metals, Y-Fe-garnets resonators [15], ceramics [19,20]
Scandium	22	NA	400 kg primary production, 1600 kg/year from Russian stockpile	Aluminium-scandium alloys for aerospace industry [21], defence industry, high intensity discharge light
Copper	60–70	1,615,000,000	17,000,000	Electrical wiring, heat exchangers, piping and roof construction and increasingly consumer electronics [6,18]

According to the Unites States Geological Survey (USGS), world resources are enough to meet foreseeable demand but world production falls short of meeting current demand [2]. The production tonnage and resources for individual rare earths are presented in Table 1. Copper, a commonly used metal, has been shown as a comparison in the last row of this table. The resources have been calculated using data on the percentage of rare earths found in various ore deposits [2] and the known resources of rare earth containing ores [15]. The only rare earth element estimated to have less than 1000 years resource is Eu at approximately 600 years. However growth in consumption of particularly the heavy rare earth elements has grown by many orders of magnitude in the past two decades, and another order of magnitude increase in demand is not out of the question since Eu is used in red phosphors (along with Y) for low energy consuming lighting suggesting that peak Eu might be as little as 10 to 30 years away. This further underlines the urgent case for recycling [16]. Although this assessment is too simplistic since the life of particular resources will be influenced by the finding of new deposits, technological efficiency such as use of less specific material per product, extraction efficiency of low grade ore, and finally potentially the stream of recovered metal from recycling. However, it is useful to identify those materials based on these potential critical factors.

Australia potentially has a role in world supply, with relatively modest deposits of REEs (2 Mt REE and yttrium oxide as Economically Demonstrated Resources (EDR), of which 31% EDR is accessible). World resource is reported as 95 Mt [22]. This Australian resource estimate excludes the tailings heap of Olympic Dam which can be a significant source of light REEs at a very low grade. The estimated demand for REEs was about 125,000 t in 2010 with 95% supply from China [23]. The reported current supply from China is estimated to be 94% [24]. The expected demand of rare earth oxide is about 200,000 t in 2014 [11] although this figure does not discriminate between which rare earths will be in highest demand. The supply was 115,000 t in 2012 and 108,000 t in 2013. Another estimate predicted that in 2016 the forecast supply would 195,000 tonnes against a demand of 180,000 t. This may not happen in reality and may have uncertainty. Clearly, further information is required in a breakdown of consumption patterns for all of these elements.

3.2. Mineralogy

Rare earth containing minerals are usually dominated by either HREEs or LREEs. Minerals containing predominantly yttrium and the HREEs include gadolinite, xenotime, samarskite, euxenite, fergusonite, yttrotantalite, yttrotungstite, yttrialite. Minerals containing predominantly LREEs include bastnasite, monazite, allanite, loparite, ancylite, parasite, lanthanite, chevinite, cerite, stillwellite, britholite, fluocerite and cerianite. However, commercially operating mines around the world mainly extract bastnasite, monazite and xenotime ores.

The major mineral deposits in Australia principally contain monazite and some xenotime whereas the US and Chinese deposits principally contain bastnasite. China and Malaysia have xenotime deposits. Some of Australia's deposits occur near Alice Springs (Arafura, Perth, Australia), Dubbo (Alkane, Perth, Australia; Jervois Mining, Melbourne, Australia), Laverton, WA (Lynas, Sydney, Australia), Nyngan and Young, NSW (Jervois Mining, Melbourne,

Australia, Sc only), Southeast Kimberley (Navigator resources Ltd, Perth, Australia), Capital (near Canberra), Gifford Creek, WA (Artemis Resources Ltd, Sydney, Australia), Mount Gee SA (Marathon Resources Ltd, Adelaide, Australia), Olympic Dam SA (BHP Billiton, Perth, Australia), Brown's Range WA & NT (Hastings Metals, Sydney, Australia & Northern Minerals, Perth, Australia), Mary Kathleen Queensland (BHP Billiton, Perth, Australia) and Greenvale Qld (Metallica Minerals Ltd, Brisbane, Australia). These deposits have a large variation in composition [25]. Although the number of identified deposits in the world may be over 850, the actual operating mines are only handful. Prominent currently operating mines are Bayan Obo in China, Mountain Pass in the US and recently opened Mount Weld in Australia. Distribution of rare earth elements in various mines is shown in Figure 1.

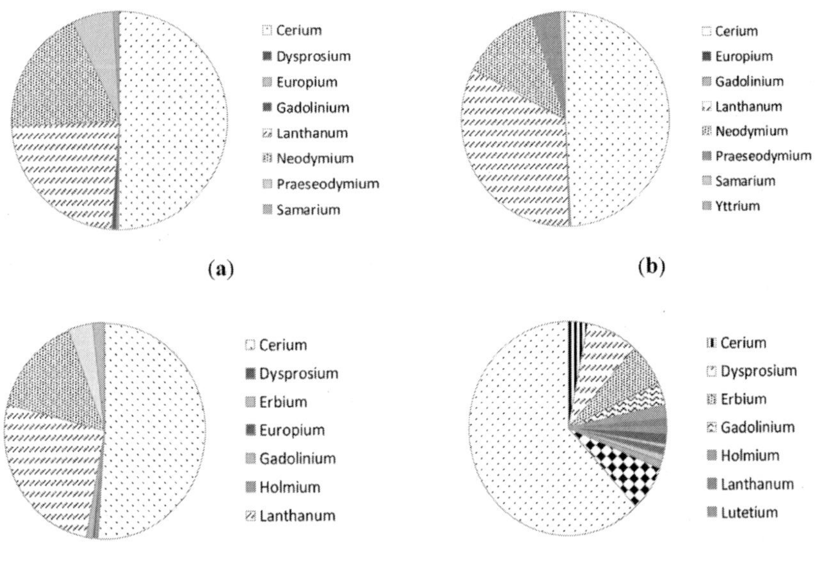

Figure 1. Average distribution of rare earth elements in each deposit [18]. (**a**) Bayan Obo, China; (**b**) Mountain pass, USA; (**c**) Mount Weld, Australia; (**d**) Lahat, Malyasia.

REEs are also often found associated with uranium/thorium mineralization and uranium ores often contain appreciable REE (and vice versa). This co-deposition with radio-nuclides can pose particular challenges in the processing of REE ores [17,22,26]. Current issues surrounding Lynas Corporation's REE plant in Malaysia relate to the perception that the plant will produce a radioactive waste. However, according to International Atomic Energy Agency, the radiation generated from the Malaysian plant would be the effect below harmful levels [27].

Mining of rare earths is generally divided into three historic eras: (i) monazite-placer; (ii) Mountain Pass and (iii) Chinese eras [26]. The advent of the Chinese era (mid 1980s) was marked by availability of rare earths at prices that undercut most other mining operations, resulting in closure of many mines outside of China. At this point, the Chinese have around 55% of all known rare earth deposits and control 95% of world supply through integrated mining, refining and supply chains.

All of these minerals are found in land based deposits, however, marine deposits have also been considered. It was reported that river run-off was considered the most dominant source of rare earth elements in the ocean [28]. However, the work [29] suggests that significant ocean rare earth sediments may arise from hydrothermal plumes from the East Pacific Rise and the Juan de Fuca Ridge. These deposits have similar to or greater levels of REE compared to the Chinese deposits and perhaps greater HREE content.

3.3. Mining Methods

Rare earth mining can be open pit, underground or leached in-situ. For a typical open pit mine, the approach is very similar to other mining operations which involve removal of overburden, mining, milling, crushing and grinding, separation or concentration. The product of the enriched concentrate after separation may contain around 30%–70% of rare earth bearing ore. The process requires higher amount of water and energy usage (e.g., compared with other common metals, *i.e.*, 0.2 to 1 GJ energy/t REO, 0.3 to 1.8 ML water/t REO) as well as production of waste streams (*i.e.*, with other metals due very low grade) such as tailings and wastewater.

If the deposit type is hard rock based then conventional open-cut or underground truck shovel mining system is used. On the other hand, if it is mineral sand based monazite type deposit then wet-dredging or dry mining method is used. If a wet mining operation, a floating dredge cuts the ore under the surface of a pond and pumps the ore slurry to a floating wet concentrator. Dry mining can be similar to conventional truck and shovel system.

3.4. Extraction and Concentration

The following description of rare earth processing is necessarily very general and represents a limited range of options that could be used in the industry. The actual process flowsheets are quite varied and tailored to the gangue and host ores such as monazite based or bastnasite based. There will be implications on impact as a result of these differences; however, this was assumed beyond the scope of this study. The extraction and processing has been described in detail [30].

To extract rare earths, further processing/extraction and refining are required. The extraction may involve using acidic or alkaline routes depending on the mineralogy of the REE-containing phases and reactivity of gangue phases. Typically, the acidic route is the most common, dominating at least 90% of the extraction methods. Depending strongly on mineralogy, the extraction step often involves roasting of the rare earth ore at 400 °C–500 °C in concentrated sulphuric acid to remove fluoride and CO_2, and to change the mineral phase to make it more water-soluble. The monazite ore processing would have different routes. The resulting ore paste is washed (usually using water) and filtered or decanted to remove fine solid impurities. The REEs are then further leached (sometimes in multiple steps) using extraction agents (hydrochloric acid) and precipitating agents (ammonium bicarbonate $(NH_4)HCO_3$ or NaOH precipitation). Further separation stages are required through, for example successive solvent extraction (e.g., ($C_{16}H_{35}O_3P$) and HCl) and then followed by precipitation steps using ammonium bicarbonate $(NH_4)HCO_3$ or oxalic acid ($C_2H_2O_4$). The precipitate is heated to form rare earth oxides (REO). LREEs may be extracted by molten salt electrolysis based on chlorides or oxides. Metallothermic reduction processes are used to extract the middle and heavy rare metals such as Sm, Eu, Tb and Dy in near vacuum conditions with inert gas at high temperatures (>1000 °C). The difficulty and complications associated with preparing commercial grade rare earth compounds and metals is more akin to fine chemical production than commodity manufacture. Additionally, valuable rare earth ores are often "incidental" and occur in association with valuable elements such as Nb, Ta, In and Ni. The flowsheets for production can therefore be further complicated with extraction steps for other valuable metals.

In an alkaline dissolution based production process, monazite concentrate is decomposed with sodium hydroxide to produce rare earth hydroxides. The rare earth hydroxides are then leached using hydrochloric acid and a mixed rare earth chloride solution is produced. The mixed solution is separated into light, medium and heavy rare earths using solvent extraction processes. Further solvent extraction stages are then required to separate these streams into solutions of individual rare earth chlorides. The separated rare earth chlorides are treated with oxalic acid to produce a rare earth hydrate, and these are then calcined in a furnace at 1000 °C to produce the final rare earth oxide products of greater than 99% purity.

In China, the rare earth elements are also recovered as a by-product of iron mining. The world's largest light rare earth deposit is Bayon Obo located in Baotou, China, containing 48 million tonnes of rare earths reserves in the form of bastnasite ore [22,24]. Bastnasite ore concentrate is typically calcined to drive off CO_2 and fluorine. It is then hydrometallurgically treated by two stage digestion with hydrochloric acid, then treatment with sodium hydroxide, respectively, to produce rare earth hydroxides. The rare earth hydroxides are chlorinated and converted to rare earth chloride products. Separation of mixed rare earth oxides into the individual rare earths can partially be achieved by multiple recrystallisation as double salts, and ultimately by multi-stage solvent extraction. The mining and processing steps for refining of rare earths, therefore, tend to be energy, water and chemical intensive with significant environment risks affecting water discharges (radionuclides, mainly thorium and uranium; heavy metals; acids; fluorides), tailing management and air emissions (radio-nuclides, Th and U, heavy metals, HF, HCl, SO_2 and dust).

3.5. Commercial REE Mining in Australia

In Australia, the principal rare earth mineral exploited has been monazite, which typically has associated radioactivity due to thorium content (by substitution up to 30%). Until 1995, rare earth production in Australia was largely as a by-product of processing monazite contained in heavy mineral sands [24]. The process involved concentrating monazite using wet processing followed by dry concentration techniques. Wet concentration separates the heavy minerals from gangue minerals. Dry concentration, such as magnetic, electrostatic and gravity separation steps are used to separate monazite from the other heavy minerals. The rare earth elements may be dissolved from monazite by high temperature leaching in concentrated sulphuric acid [31].

In 2007, mining began at the Mount Weld deposit, Laverton, WA in Western Australia, which is the richest known deposit of rare earths in the world. This deposit's mineralogy has been described as a secondary rare earth phosphate, but those phosphates (most likely monazite) are encapsulated in iron oxide minerals. Parts of the deposit have distinct xenotime mineralogy, with heavier rare earth components. The first crushed ore was fed to the concentration plant in 2011. Mount Weld is claimed to have the highest grade known deposit of rare earths in the world. This deposit consists of the Central Lanthanide Deposit ("CLD") and Duncan Deposit. The mine has a conventional open-pit operation. About 773 kt ore has been mined at 15.4% REO grade or 116 kt contained REO. The stockpiled ore is sufficient to sustain steady state production for 6 years for Phase 1 capacity of concentration plant [32].

The concentrator commissioned in 2011 is designed to process 121,000 tonnes of ore per year producing 33,000 tonnes of concentrate in the first phase using flotation technology. The second phase is expected to double this production. It is intended that the concentrates will be exported to the Lynas advanced materials plant (LAMP) built in Malaysia. As reported in June 2013, 15,000 t of dry bagged concentrate has been made ready for shipment to LAMP [32].

The LAMP is located in the Gebeng Industrial Estate, in Pahang. Initial capacity in 2012 is designed at 11,000 tonnes of separated REOs per year, with expansion to 22,000 tonnes per year in 2013 [33]. While the specific details of the flowsheet for processing concentrate are not available, they are likely to include steps of calcination, caustic conversion, acid leach and solvent extraction. Lynas reported some issues as identified relating to clogging and premature wearing of equipment that are affecting the ability to operate continuously at nameplate production capacity of LAMP. A series of work programs involving equipment changes and materials handling was implemented by Lynas in the late 2013 [32]. Their immediate target is to optimise production at the Phase 1 capacity level of 11,000 tpa REO until market prices recover. In the June 2014 quarter, the company produced 1,882 tonnes of rare earth oxides (REO), up 73% over the previous quarter. The full year production to date in September 2014 was 3,965 tonnes [33]. Lynas claimed in June 2014 [32] in their corporate website that the production rate is on track to achieve the target.

The local residents had concerns over the operation of LAMP rare-earth processing plant particularly on the contamination of the coastal environment and the adverse health impact which could result from the mismanagement of radioactive waste streams. Recently, the Atomic Energy Licensing Board (AELB) of Malaysia agreed to issue a two-year full operating stage license to Lynas after it fulfilled all the conditions [34]. The AELB assured residents that it was monitoring the operations of the Lynas plant and found radiation levels onsite and offsite to be within acceptable limits.

The reported mine production of rare-earth oxides in Australia, including yttrium oxide, was estimated to be 2,070 tonnes in 2011 and 4,000 tons in 2012 [18]. However, the Australian production of REO and Y_2O_3 is reported as 2,070 kt [1] in 2011 possibly due to a unit error. This should be 2,070 tonne, not in kilo-tonne. Unfortunately, there is a lot of anomaly in the reported production, resources, reserve and life of resources reported by various sources that is extremely difficult to verify particularly for REEs.

3.6. REE Mining in China

Between 2010 and 2012 the Chinese government put strict export quotas on their rare earth minerals and semi processed rare earth products [2,17,35,36]. The quotas reduced the output by nearly 60% compared to the 2008 total release of 34,156 tonnes [2]. These quotas created a gap between demand and supply and large increases in the prices of the rare earths. In relation to a dispute under World Trade Organisation, the United States, European Union, Canada and Japan requested consultations with China with respect to China's restrictions on the export of various forms of rare earths, tungsten and molybdenum in 2012. At the recent Dispute Settlement Body (DSB) meeting on 26 September 2014, China stated that it intended to implement the DSB's recommendations and ruling in a manner that respects its WTO obligations. China added that it would need a reasonable period of time to do so [37].

Because rare earths are not traded in a commodity market such as the London Metal Exchange, accurate records of prices are difficult to establish. However, from this source [38] it is possible to establish that dramatic increases in prices up to 10 fold were observed over the decade. The prices of several oxides are presented in Figure 2 and Figure 3 that show the significant volatility of rare earth market. One example includes neodymium oxide which increased from just under $10/kg in 2001 to nearly $239/kg in 2011 but came down to $80/kg, which is about 3 times reduction from the peak. Other price increases include Europium, Dysprosium and Terbium oxide in the order of 300%–500% increase over the last 10 years [39]. Recent trends in 2012 suggest that rare earth prices have decreased from their maxima in 2011 and are beginning to stabilise, albeit at significantly higher levels than prices even in 2010 for some with upward trend (Dy, Pr, Gd, Eu, Nd), no change (Tb) but reduction for other oxides (La, Ce, Sm). Announcements of new rare earth reserves in North Korea and Canada will also put downward pressure on prices.

The Chinese producers also have issues with environmental pollution resulting from poorly regulated mining operations, as well as reported smuggling of rare earths from the Southern provinces which have been estimated to account for 30% of total Chinese production [19,35].

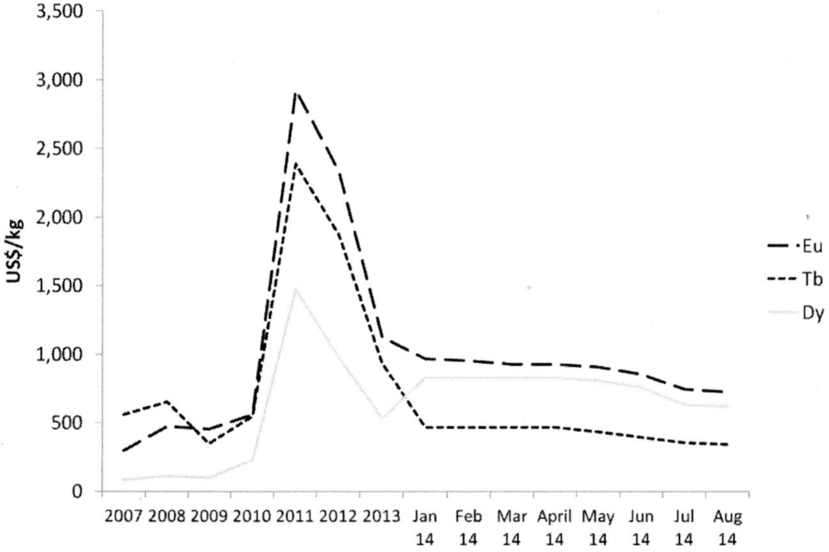

Figure 2. Price history in US dollars of the expensive rare earths [40].

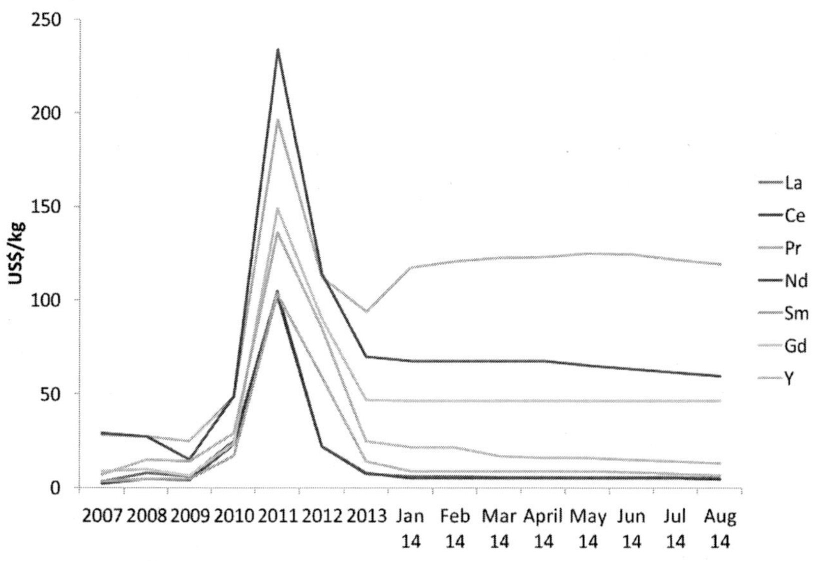

Figure 3. Price history in US dollars of the comparatively cheaper rare earths [40].

3.7. RRE in Korea

A major impact on the rare earth market occurred with the announcement of rare earth deposits at Jongju in North Korea which are being developed by Pacific Century Rare Earth Mineral and the North Korea government. These reserves double the current estimates of rare earths reserves. Moreover, the rare earth deposits are relatively rich with 664.9 Mt at 9% TREO, 634 Mt between 5.7% and 9.0% and another couple of billion tonnes between 3.97% and 5.7% along with another 2.8 billion tonnes of lower grade ores. The total amount of lower grade REO is estimated to be around 216.2 Mt. Further exploratory drilling is expected to be undertaken this year and reported when the drilling is finished. However, veracity of this announcement is questioned by the sources of industry intelligence. Furthermore, there is a complication of making this project into reality due to political implications and international sanctions in the short to medium term.

3.8. RRE in Canada

Until the discovery of the Jongju deposit(s) the Canadian deposit at Nechalacho in the North-West Territories was one of the most exciting rare earth developments in recent years. This large deposit is rich in heavy rare earth elements consists of the following minerals: LREE in bastnaesite, synchisite, monazite and allanite and HREE in zircon, fergusonite and rare xenotime. The mine is fully owned by Avalon Rare Earth metals. Avalon also has a number of other sites in North America. The mine site treats the ores using hydrometallurgical approaches with acid leaching as the key processing step [41]. The % of HREE to total RE varies from 6% to 30% from the top to bottom of the deposit. Total deposit is estimated to be 1.71 Mt of rare earth oxide. As with most REE mining thorium and uranium is associated with the deposit. Production is to begin in late 2016.

3.9. Uses

Some of the largest uses of rare earths are in catalysts (20%, largely Ce and La), rare earth magnets (21%, largely Nd, Sm and

Dy), alloys (18%), powder production (12%) and phosphors (7%) [42]. Catalyst applications are for both industrial and auto catalysts. Phosphors are important for a range of applications particularly for visual display in screens and low energy lighting [26]. This is a likely growth area and will put pressure particularly on Eu and Tb reserves. Another area for expansion will be rare earth magnets (Nd, Pr, Sm and Dy) particularly for alternative energies. These will find widespread application in wind turbines, the auto industry (electric and hybrid cars) and defence industry (*i.e.*, missile guidance systems). Rare earth containing (Er) glasses are important for fibre optical amplifiers required in high speed optical communication networks [26,42]. Improving the efficiency of solar energy conversion is another area of probable expansion for rare earths [42,43,44]. The rare earths Er, Yt and Ho show promise for up-conversion (converting low energy photons into higher energy photons by increasing the wavelength *i.e.*, infra red photons to visible) and Yb, Tm and Tb for down-conversion (converting high energy photons to lower energy) [43,44].

Some novel areas for rare earths include refrigeration using either laser cooling or the magnetocaloric effect [42]. In laser cooling, vibrational energy is removed by emission of photons with a higher average energy than the absorbing photons in various materials that may include Y, Yb, La, Nd and Tm [45,46]. In the magnetocaloric effect, a material has its magnetic domains aligned under a strong magnetic field and when the field is removed the domains can unalign using phonons energy causing cooling. Room temperature magneto-caloric materials (e.g., Gd5(SixGe(1-x)) [46] may offer an alternative to current vapour compression cycles used with current refrigerants [47]. Some more mundane uses include flints for cigarette lighters (Ce), polishing agents (CeO_2), rechargeable batteries (La) and carbon arc lamps (La) [26].

Growth in demand for these materials is certainly of concern in many countries which is best summarised in some US documents [15,26]. Rare earths underpin technologies that are seen as critical for clean energy economies, such as in photovoltaic devices, battery technologies for transport and wind energy, and phosphors for lighting. High speed optical fibre communication is another likely growth area. In fact a risk analysis undertaken by the U.S. Department of Energy nominates Dysprosium,

Neodymium, Terbium, Europium and Yttrium as critical to advancement in a clean energy future [48].

3.10. Recycling

Unlike other metal recycling industry, the recycling industry for rare earth is evolving with many of the activities still at development stages. The need for recycling is in part driven by rapid increase in demand and scarcity of materials combined with security of supply issues, a problem exacerbated by additional export restrictions imposed by the Chinese authority. The recycling option, however, offers several advantages including lower environmental impact, less dependency on Chinese export and the feed materials can be free from radioactive contaminants (e.g., Th and U).

Possible opportunities available for recovery of rare earths are typically from used magnets, batteries, lighting and spent catalysts. The materials could be sourced from post-production wastes and post-consumer disposed goods. Clearly, the latter option is expected to be more complex, requiring more extensive physical and chemical treatments and higher energy requirement.

Typical post-production wastes may, for example, include scraps produced during manufacturing of magnets (with expected wastage of 20%–30%) or sludge produced from shaping and grinding of magnet alloys [49]. Typical recovery methods may include re-melting, reuse and re-blending of scraps or other selective extraction methods (e.g. using a molten salt agent ($MgCl_2$), Na_2SO_4 double-salt precipitation technique or oxalate secondary precipitation approach for recovery of Nd_2O_3 [50]. There are already metal recyclers that have identified the value in magnet-manufacturing dross, with pilot programs using standard and simplified hydrometallurgical routes to recover Nd and Dy.

Recycling of rare earths from consumer goods at end-of-life is generally more challenging due to dispersed nature of the rare earth compounds which are intricately embedded into the products (e.g., neodymium magnets from hard disks and compressors, electronic displays). For effective recycling, it requires efficient and effective physical and chemical separation techniques. Physical separation methods may include mechanical dismantling of the components such as that proposed by Hitachi in which the neodymium magnets from hard disks and compressors motors/generators are selectively removed by a purpose designed automated machine, enabling significantly faster (8 times) separation than manual labour. Such a recycling approach is expected to meet 10% of Hitachi's needs by 2013 [51]. Other recovery approaches have been outlined for magnets from MRI (magnetic resonance imaging for medical application) [49] and fluorescent lamps and tubes [52]. A potential future strategy for industry is to collaborate on the design of RE magnet components to achieve some degree of dimensional standardisation and therefore reusability, and industrial design with easy recycling as an objective.

The chemical recovery methods for rare earth elements generally involve pyro-metallurgical and hydrometallurgical approaches [53,54]. Pyrometallurgical routes are commonly used for recovery of metals from electronic scraps. However, the rare earths are easily lost using this method as they tend to report to the slag phase due to the high affinity of the rare earth elements to oxygen. Hence hydrometallurgical approaches may be required to leach the rare earth elements from the slag. Rare earth elements (e.g., lanthanum, cerium, neodymium and praseodymium) and other valuable metals (Ni, Co) can be recovered from used Ni-MH batteries by leaching with sulphuric acid [55]. While for spent catalysts, it is still not yet considered technically feasible and economically acceptable for the recycling of rare earth elements (mainly La and Ce) from FCC catalysts.

The losses of the consumer goods through exportation and uncontrolled disposal are thus making the recovery of resources difficult. One of the primary constraints is to have effective collection system and consumer cooperation in recycling of end-of-life devices. Although the R&D in this domain is gaining momentum but there is still a lack of fully developed commercial scale plants for rare earth recycling in part due to drawbacks on yields and cost. In addition, some of the devices containing rare earths are still new in the market place and it would take some time for the products to reach the end of their life (10–20 years for electric motors in vehicles and wind turbines) before they are available for recycling.

Another approach to address the rare earth supply issues is substitution with materials made from more common elements but this approach is still in its early stages. Possible opportunities for substitution under consideration are alternative turbine types for wind turbines; improving the reliability of traditional turbines with gears and alternative motor designs motors of hybrid and fully electric vehicles. However, it is conceived that substitution opportunities in applications related to energy efficient lighting systems containing rare earths (compact fluorescent lamp, LED, plasma display, LCD display); or catalysts in automotive and petroleum refining are still difficult and unlikely in the short term horizon without further research.

3.11. Dispersion and Issues with Recovery

As with all scarce elements one of the biggest issues is dispersion at the end-of-life for products into which these elements are incorporated. In the past most discarded product and waste from processing has ended up in landfill or dumped in the ocean. While there is discussion about mining waste dumps, material disposed of at sea is effectively lost, *i.e.*, dispersed into the environment. In the context of rare earths, most products where the more precious rare earths are used are not inherently dispersive and programs exist or are beginning in many countries for reclamation and driven by recent increases in the price of rare earths [56,57,58].

Potential dispersive uses of rare earths include catalysts, magnets (Sm, Nd), missiles (Sm, Nd,) sunglasses (Er), MRI agents (Gd), lighting (Er, Eu, Tb) and cigarette lighters (Ce). Industrial catalysts generally go to landfill and can potentially be recovered from landfill sites in the future if they are not dispersed. In Europe there is a tendency to dispose of these catalysts into cement products as clinker, which is obviously a loss. While a large proportion of any automotive catalyst can, in principle, be recovered at end-of-life, degradation during the operational vehicle lifetime results in the formation of dust and dispersion of the dust onto roadways during exhaust gas flow. The dispersed material ends up in river systems as a result of storm water flow and effectively dispersed back into the environment. For automotive catalysts, this equates to the dispersive loss of the active metal catalyst (usually Pt is more important than that of CeO_2 which is also an active ingredient of the catalyst). Products including high field strength rare earth magnets are another potential source for dispersion. This is an interesting category as products with miniaturised electronics such as mobile phones tend not to be discarded but stored in Australian households [59] and potentially available for rare earth recovery. However, the potential for mobile phones to be collected has also been documented [60]. This behaviour is not unique to either electronic goods or Australian households but common across the OECD [61].

On the other hand rare earth magnets used in missile guidance systems will be dispersed into the environment if the missile is deployed and used. Lighting products have traditionally been sent to landfill at end of life, but programs are now being considered for recovery of the rare earths in these products as discussed in the "Recycling" Section. Cigarette lighters (rare earth flints) are generally dispersed into the environment, either in landfill or storm water drains. Erbium used as fibre optic amplifiers for high speed communications might also end up being dispersed, in this case because the Er is used at dopant level [62] making it expensive to recover.

The rare earths at greatest risk are those that have high volume, low reserves and significant dispersion. Essentially, rare earth elements are concentrated from very low grade in the ground and then dispersed into the various equipments in small quantities. Thus, unless recovered, there is greater chance of these materials lost with the disposal of the equipment.

3.12. Life Cycle Based Environmental Impact (Demonstration of One Data Set)

For life cycle based environmental impact assessment, one simple case study has been selected from the Ecoinvent database of SimaPro LCA software [63,64]. The mining and concentration of a bastnasite ore with a rare earth oxide concentration of 6% has been assumed in China [65]. The boundary includes the mining, concentration of REE oxides and separation of products such cerium concentrate (60% cerium oxide), lanthanum oxide, neodymium oxide, praseodymium oxide, and samarium europium gadolinium (mixed medium heavy REE concentrate, 94% rare earth oxide). The global warming potential impact or greenhouse gas (GHG) footprint of rare earth production has been estimated and reported in kg CO_2-equivalent (kg CO_2-e) unit per kg of REO production.

The process included material and energy inputs, emissions and land use for the mining and concentration of a bastnasite ore with a rare earth oxide concentration of 6%. Input and outputs were reported from bastnasite ore composition mined in China. Infrastructure and land use was approximated with iron ore mining, assuming Chinese Bayan Obo mine which is largely started as an iron ore mine (Fe content is about 33%) but now produces over 95% of Chinese and 83% of global REE production. Some inputs of auxiliary materials were estimated according to stoichiometry. The estimation for energy consumption, wastes and emissions is indicative only. Production in China was considered for this study. This process may be applicable for other regions if a similar ore type is used.

The assumed subsequent process produced products such cerium concentrate (60% cerium oxide), lanthanum oxide, neodymium oxide, praseodymium oxide, and samarium, europium, gadolinium (mixed medium heavy REE concentrate product with 94% rare earth oxide). This process included roasting and cracking of the REE concentrate with 98% sulphuric acid at 500°C in a rotary kiln. Solvent extraction (SX) was used for the separation of the different rare earth oxides where organic chemicals were used. The obtained rare earth oxide product has a purity of up to 99.9%. The revenue from each product was used to allocate environmental burden. The break-down of various energy and materials on the total greenhouse gas (GHG) emission from the production process of REE concentrate is shown in Figure 4.

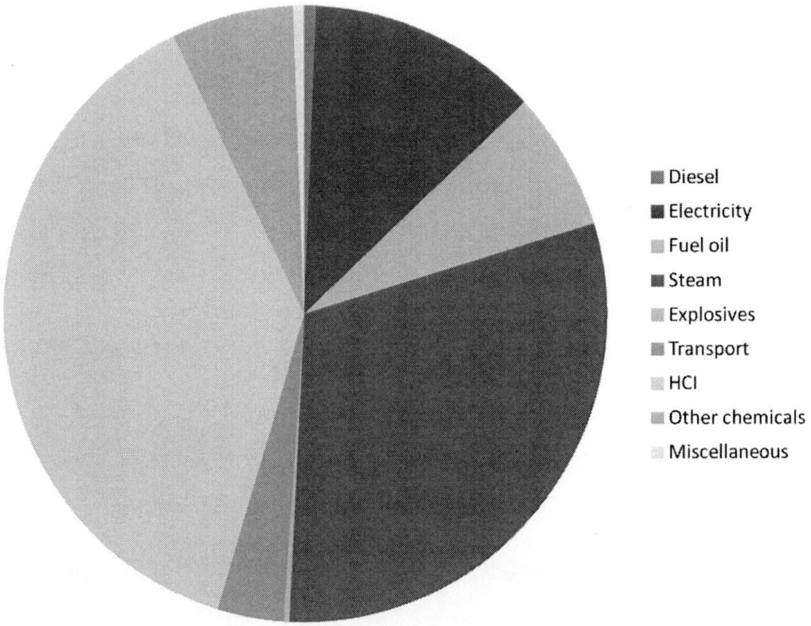

Figure 4. Distribution of impact for producing mixed rare earth elements (REE) oxides (total 1.4 kg CO_2-e/kg REE oxides).

The major contributor to total GHG footprint of REE processing is hydrochloric acid (*ca.* 38%), followed by steam use (32%) and electricity (12%). Overall 51% of GHG is due to use of energy in various forms (*i.e.*, diesel, steam, fuel oil and electricity). The remaining minor elements are from other chemicals and transport. To reduce GHG impact of REE processing, the focus should be on how to reduce the acid and energy consumption during processing. The reported GHG is an order of magnitude higher in a study published recently [66]. There is always difference between reported results from an LCA due a variety of reasons that include the assumed boundary and allocation of impact. Ore grade is one of those main ones, which is assumed as 4% for their study. There will be differences on environmental impact if different ore type, deposit or mining methods are assumed. In this paper, although GHG impact has been selected, other impacts such as radio-activity will be important which has been considered beyond the scope of this present paper.

The GHG footprint of separated oxides of RE elements from the respective materials and energy contribution is shown in Figure 5.

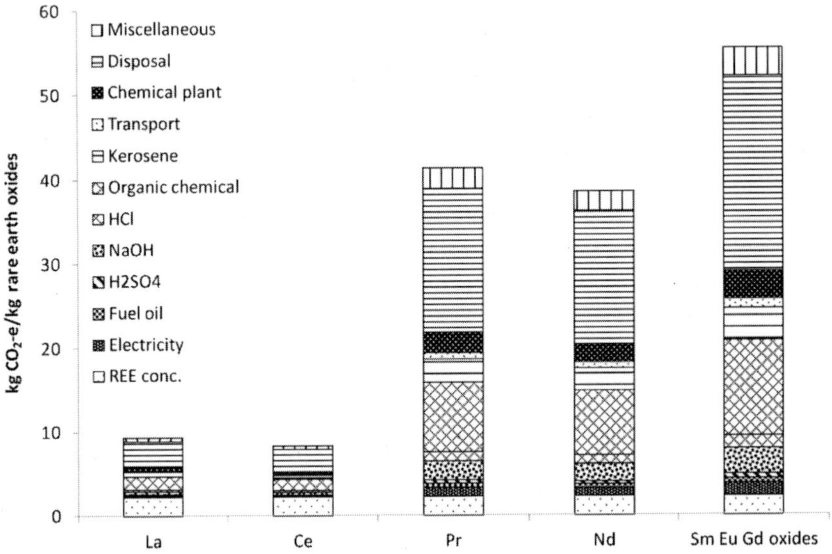

Figure 5. Greenhouse gas footprint of selected rare earth products (note: disposal includes hazardous solvent incineration, 70% mixed REE is processed, 90% recovery to products, mine rehabilitation is under miscellaneous [67].

Disposal steps of solvents mixture with water and tailings and to hazardous waste incineration contributed significant amount of GHG emission for each separated RE oxide product. Use of organic chemical (during solvent extraction) had the next largest contribution on the total GHG footprint. Thus, the effort should focus on the reduction of the use of organic chemicals during rare earth purification and on the disposal issues.

Table 2 shows the few selected life cycle based environmental impact of selected rare earth elements production (Australian indicator set method was chosen for this impact analysis).

Table 2. Environmental footprint of selected rare earth oxides production.

REOs	Primary Energy MJ/kg	GHG kg CO_{2-e}/kg	Water kL Water/kg	Toxicity *DALY/kg
La	177	9.3	0.33	1.65×10^{6}
Ce	157	8.3	0.30	1.46×10^{6}
Pr	798	41.4	1.32	7.36×10^{6}
Nd	743	38.5	1.23	6.86×10^{6}
Sm, Eu, Gd (mixed oxide)	1,074	55.6	1.75	9.89×10^{6}

Note: *DALY—Disability adjusted life years (metric to determine toxicity on human health developed by World Health Orgnisation [64].

The environmental footprints depend on the ore grade and recovery of the particular REEs. For example, La and Ce is generally available in higher amounts in ore and concentrate and thus can be recovered more readily compared with other REEs. Thus their (La and Ce oxides) specific impact is relatively less compared with other elements. Since the energy footprint of REE oxides are similar to that of other metals, the footprints of REE metals would significantly be higher (with further processing stages) even with energy intensive metals such as aluminium (211 MJ/kg) and titanium (361 MJ/kg) [68]. Their water footprints are much higher than that of the results for most metal studied in the past (*i.e.*, titanium water footprint is reported to be 110 kL/t metal [69]).

The environmental impacts such as radioactivity potential, acidification, eutrophication, solid waste generation, water use, gross primary energy footprint, toxicity and any other impact of significance on regional and global basis should be taken into consideration. However, most of the REEs are expected to be used in energy reduction, energy efficiency and renewable energy technologies. Thus future LCA studies should include the boundary of the use phase and their impacts on overall emission reduction that requires further data and analysis that is beyond the scope of this current study.

4. CONCLUSIONS

The availability of rare earths is in transition from a temporary decline due mainly to quotas being imposed by the Chinese government on export to becoming more available with production increasing elsewhere in the world. The reduction in availability coupled with increasing demand led to increased prices for rare earths in 2012 and 2013, but prices have fallen from their peak values. The increasing demand for rare earths in a range of applications means that the rare earth market is likely to be demand-driven for some time to come. This is particularly the case for rare earths used in high field strength magnets. Price fluctuations are therefore likely to vary until such time as new rare earth deposits supply the market or formerly closed mines are reopened.

The rare earths at greatest risk are those that have high volume, low reserves and significant dispersion. Essentially, rare earth elements are concentrated from very low grade in the ground and then dispersed into various equipments in small quantities. Thus, unless recovered, there is greater chance of these materials lost with the disposal of the equipment. Recycling and recovery of rare earth poses challenge in terms use of energy to collection, reprocessing and reproducing products at specification that can replace primary metals.

The major contributor to total GHG footprint of REE processing is hydrochloric acid (*ca.* 38%), followed by steam use (32%) and electricity (12%). Overall, 51% of GHG is due to use of energy in various forms (*i.e.*, diesel, steam, fuel oil and electricity). The remaining minor elements are from other chemicals and transport. To reduce GHG impact of REE processing, the focus should be on how to reduce the acid and energy consumption during processing.

ACKNOWLEDGMENTS

The authors are thankful to internal reviewers (Stephen Northey, Andreas Monch) in CSIRO and external anonymous reviewers and for their comments to improve this paper.

AUTHORS CONTRIBUTIONS

Nawshad Haque collected, analysed literature, and prepared the manuscript for submission. He also contributed the environmental impact section and coordinated overall aspects of the paper until its publication.

Anthony Hughes collected literature, analysed and prepared the first draft of the manuscript. Anthony Hughes assisted to address reviewer's comments and contributed to prepare and finalise published version.

Seng Lim collected literature, assisted to prepare the first draft, reviewed and commented for improvement of final published version.

Chris Vernon collected literature, assisted to prepare the first draft, reviewed and commented for improvement of final version and contributed to prepare the published version.

CONFLICTS OF INTEREST

The authors declare no conflict of interest.

REFERENCES

1. Skirrow, R.G.; Huston, D.L.; Mernagh, T.P.; Thorne, J.P.; Dulfer, H.; Senior, A.B. *Critical Metals for a High-Tech World: Australia's Potential to Supply Global Demand*; Geoscience Australia: Canberra, NSW, Australia, 2013; pp. 10–13.
2. Cordier, D.J.; Hedrick, J.B. Rare earths: U.S. Geological Survey Minerals Yearbook 2008. Available online: http://minerals.usgs.gov/minerals/pubs/commodity/rare_earths/myb1-2008-raree.pdf (accessed on 27 October 2014).
3. Tharumarajah, A.; Koltun, P. Cradle to gate assessment of environmental impact of rare earth metals. In Proceedings

of the 7th Australasian Life cycle assessment Conference, Melbourne, Australia, 9–10 March 2011.

4. Foley, J. Boundaries for healthy planet. *Sci. Am.* **2010**, *302*, 54–57.

5. Cordell, D.; Drangert, J.-O.; White, S. The story of phosphorus: Global food security and food for thought. *Glob. Environ. Chall.* **2009**, *19*, 292–305.

6. Memary, R.; Giurco, D.; Mudd, G.; Mohr, S.; Weng, Z. *Copper Case Study: Australian Resources, Technology and Future Scenarios*; Prepared for CSIRO Minerals Down Under Flagship; Institute for Sustainable Futures, University of Technology, Sydney: Sydney, Australia; the Department of Civil Engineering, Monash University: Sydney, Australia, 2012; pp. 10–20.

7. Northey, S.; Mohr, S.; Mudd, G.M.; Weng, Z.; Guirco, D. Modelling future copper resources and mining. *Resour. Conserv. Recycl.* **2014**, *83*, 190–201.

8. Alonso, E.; Sherman, A.M.; Wallington, T.J.; Everson, M.P.; Field, F.R.; Roth, R.; Kirchain, R.E. Evaluating rare earth element availability: A case with revolutionary demand from clean technologies. *Environ. Sci. Technol.* **2012**, *46*, 3406–3414.

9. Binnemans, K.; Jones, P.T.; Blanpain, B.; Van-Gerven, T.; Yang, Y.; Walton, A.; Buchert, M. Recycling of rare earths: A critical review. *J. Clean. Prod.* **2013**, *51*, 1–22.

10. Golev, A.; Scott, M.; Erskine, P.D.; Ali, S.H.; Ballantyne, G.R. Rare earths supply chains: Current status, constraints and opportunities. *Resour. Policy* **2014**, *41*, 52–59.

11. Weber, R.J.; Reisman, D.J. *Rare Earth Elements; A Review of Production, Processing, Recycling and Associated Environmental Issues*; US EPA Region 7: Kansas, KS, USA, 2012.

12. McLellan, B.C.; Corder, G.D.; Ali, S.H. Sustainability of rare earths—An overview of the state of knowledge. *Minerals***2013**, *3*, 304–317.

13. Mudd, G.M. Gold Mining in Australia: Linking historical trends and environmental and resource sustainability.*Environ. Sci. Policy* **2007**, *10*, 629–644.

14. Norgate, T.; Haque, N. Using life cycle assessment to evaluate some environmental impacts of gold production. *J. Clean. Prod.* **2012**, *29–30*, 53–63.

15. Humphries, M. *Rare Earth Elements: The Global Supply Chain: Congressional Research Service*; The Library of Congress: Washington, DC, USA, 2010.

16. GE, Lighting. Available online: http://www.gelighting.com (accessed on 11 September 2013).

17. Hurst, C. *China's Rare Earth Elements Industry: What Can the West Learn?* Institute for the Analysis of Global Security: Fort Leavenworth, KS, USA, 2010.

18. USGS. USGS Minerals Information. Available online: http://minerals.usgs.gov/minerals/pubs/commodity/rare_earths(accessed on 11 September 2013).

19. MMTA. Minor Metals Trade Association. Available online: http://www.mmta.co.uk/metals/Ce (accessed on 2 November 2013).

20. Hughes, A.E. Interfacial Phenomena in Y2O3-ZrO2-Based Ceramins: A Surface Science Perspective. *Mater. Sci. Monogr.***1995**, *81*, 183–238.

21. Ahmad, Z. The properties and application of scandium-reinforced aluminium. *JOM* **2003**, *55*, 35–39.

22. Geoscience Australia. Australia's Identified Mineral Resources 2012. Commonwealth of Australia, Canberra. Available online: http//www.ga.gov.au (accessed on 8 May 2013).

23. Kingsnorth, D.J. *Rare Earth Opportunities, Real or Imaginary?* IMCOA Pty Ltd.: Melbourne, Australia, 2011.

24. Geoscience Australia. Rare Earth. Available online: http://www.ga.gov.au/minerals/mineral-

resources/rare-earth-elements.html (accessed on 11 September 2013).

25. Miezitis, Y. *Rare Earths in: Australia's Identified Mineral. Resources*; Australian Government: Canberra, Australia, 2010.

26. Haxel, G.B.; Hedrick, J.B.; Orris, G.J. Rare Earth Elements—Critical Resources for High. Technology, USGS Fact Sheet 087–02. Available online: http://pubs.usgs.gov/fs/2002/fs087-02/ (accessed on 20 October 2014).

27. Heber, A. Solving a sticky situation: Mineral processing is turning green to extract rare earths at cheaper rates.*Australian Mining* **2013**, 60.

28. Dubinin, A.V. Geochemistry of rare earth elements in the ocean. *Lithol. Miner. Resour.* **2004**, *39*, 289–307.

29. Kato, Y.; Fujinaga, K.; Nakamura, K.; Takaya, Y.; Kitamura, K.; Ohta, J. Deep sea mud in the Pacific Ocean as a potential resource for rare-earth elements. *Nat. Geosci.* **2011**, *4*, 535–539.

30. Gupta, C.K.; Krishnamurthy, N. *Extractive Metallurgy of Rare Earths*; CRC Press: Boca Raton, FL, USA, 2005.

31. Hedrick, J.B. *Rare Earth Minerals and Metals 1990*; United States Bureau of Mines: Washington, DC, USA, 1992.

32. Lynas Corporation. Investor Presentation 1 August 2013. 2013. Available online: http://www.lynascorp.com/Pages/home.aspx (accessed on 16 September 2014).

33. Motley Fool Australia. Here's why Lynas Corporation Limited is getting crunched today. Available online: http://www.fool.com.au/2014/07/31/heres-why-lynas-corporation-limited-is-getting-crunched-today/# (accessed on 27 October 2014).

34. The Malaysian Insider. Lynas gets full operating licence before TOL expiry date. Available online: http://www.themalaysianinsider.com/malaysia/ar

ticle/lynas-gets-full-operating-licence-before-tol-expiry-date (accessed on 27 October 2014).

35. Xinhua. China sets first rare earth export quota for 2012. Available online: http://news.xinhuanet.com/english/china/2011–12/27/c_131329820.htm (accessed on 10 October 2013).

36. Xinhua. China to introduce special invoice for rare earth sector to curb illegal mining. Available online: http://news.xinhuanet.com/english2010/china/2011–11/30/c_131279962.htm (accessed on 10 October 2013).

37. World Trade Organization. China: Measures Related to the Exportation of Rare Earths, Tungsten and Molybdenum. Available online: http://www.wto.org/english/tratop_e/dispu_e/cases_e/ds431_e.htm (accessed on 17 October 2014).

38. Roskill. Roskill data. Available online: http://www.roskill.com (accessed on 11 September 2013).

39. Schuler, D.; Buchert, M.; Liu, R.; Dittrich, S.; Merz, C. *Study on Rare Earths and Their Recycling: Final Report for The Greens/EFA Group in the European Parliament*; Oko-Institute: Darmstadt, Germany, 2011.

40. Arafura Resource Ltd. Arafura Pricing. Available online: http://www.arafuraresources.com.au/rare-earths/pricing.html (accessed on 16 September 2014).

41. Mineral Industry Consultants. Technical report disclosing the results of the feasibility study on the Nechalacho rare earth elements project. Available online: http://www.sec.gov/Archives/edgar/data/1362898/000114420413064890/v361732_ex99-1.htm (accessed on 27 October 2014).

42. Bünzli, J.G.; Eliseeva, S.V. Lanthanide NIR luminescence for telecommunications, bioanalyses and solar energy conversion. *J. Rare Earths* **2010**, *28*, 824–842.

43. Richards, B.S. Enhancing the performance of silicon solar cells via the application of passive luminescence conversion layers. *Solar Energy Mater. Solar Cells* **2006**, *90*, 2329–2337.

44. Van der Ende, B.M.; Aarts, L.; Meijerink, A. Lanthanide ions as spectral converters for solar cells. *Phys. Chem. Chem. Phys.* **2009**, *11*, 11081–11095.

45. Ruan, X.L.; Kaviany, M. Advances in laser cooling of solids. *J. Heat Transfer-Trans. ASME* **2007**, *129*, 3–10.

46. Pecharsky, V.K.; Gschneidner, K.A. Giant magnetocaloric effect in Gd-5(Si2Ge2). *Phys. Rev. Lett.* **1997**, *78*, 4494–4497.

47. Pecharsky, V.K.; Gschneidner, K.A. Magnetocaloric effect and magnetic refrigeration. *J. Magn. Magn. Mater.* **1999**, *200*, 44–56.

48. USDE. Critical materials strategy. Available online: http://energy.gov/sites/prod/files/edg/news/documents/criticalmaterialsstrategy.pdf (accessed on 11 September 2013).

49. Shirayama, S.T. Selective extraction of Nd and Dy from rare earth magnet scrap into molten salt. In Proceedings of the Minerals, Metals & Materials Society 2009 Meeting, San Francisco, CA, USA, 15–19 February 2009.

50. Tang, J.; Wei, C.; Zhao, D.; Hong, L.; Tian, G. Nd_2O_3 Recovery from Sintered NdFeB scrap. *Rare Metals Cem. Carbides***2009**, 1–4.

51. Bloomberg. Hitachi Leads Rare Earth Recycle Report. 2010. Available online: http://www.bloomberg.com/news/2010-12-08/hitachi-recycles-rare-earth-as-china-crimps-supply.html (accessed on 11 September 2013).

52. Guarde, D.; von Heeren, C.; Sanchez, J.C.; Arnaiz, S.; Iribarnegaray, D.; Castillo, B. Improved Management and Recycling of Waste fluorescent lamps. The Prolux Project. In Proceedings of the 5th Going Green—CARE INNOVATIONS, Vienna, Austria, 8–10 November 2010.

53. Takeda, O.; Nakano, K.; Sato, Y. Resource recovery from waste of rare earth magnet by utilising fluoride molten salts.*Yooyuen Oyobi Koon Kagaku* **2009**, 63–70.

54. Koyama, S.; Suzuki, T.; Ozawa, M. From waste to resource, nuclear rare metals as a dream of modern alchemists.*Energy Convers. Manag.* **2010**, *51*, 1799–1805.

55. Wu, W.; Zhang, H. Recovery of Ni, Co, Rare Earth from Waste MH-Ni Battery. *Chin. J. Rare Metals* **2010**, *1*, 79–84.

56. Reuters. Recycling Rare earth element. Available online: http://www.reuters.com/article/2011/10/03/rare earths-rhodia-idUSL5E7L33FO20111003 (accessed on 27 October 2014).

57. Rhodia. Rhodia Group: Rhodia develops innovative process for the recycling of rare earths. Available online: http://www.rhodia.com/en/news_center/news_rel eases/Recycle_rare_earths_031011.tcm (accessed on 27 October 2014).

58. Rhodia. Rhodia Group: Umicore and Rhodia develop unique rare earth recycling process for rechargeable batteries. Available online: http://www.rhodia.com/en/news_center/news_rel eases/Umicore_rare_earth_160611.tcm (accessed on 27 October 2014).

59. Read, R. *Australia: A Nation of Hoarders*; Mobilemuster, Australian Mobile Telecommunications Association: Canberra, NSW, Australia, 2010.

60. Reller, A.; Bublies, T.; Staudinger, T.; Oswald, I.; Meissner, S.; Allen, M. The mobile phone: powerful communicator and potential metal dissipator. *GAIA* **2009**, *2*, 127–135.

61. Johnstone, N. Decision-Making and Environmental Policy Design for Consumer Durables. Available online: http://www.oecd.org/officialdocuments/publicdis playdocumentpdf/?doclanguage=en&cote=ENV/EPOC/WP NEP(2002)7/final(accessed on 21 October 2014).

62. Desurvire, E.; Simpson, J.R.; Becker, P.C. High-gain erbium-doped travelling-wave fiber amplifier. *Optics Lett.* **1987**, *12*, 888–890.

63. Ecoinvent. Ecoinvent centre, Swiss Centre for Life Cycle Inventory. Available online: http://www.ecoinvent.ch(accessed on 11 September 2013).

64. PRe. PRe Sustainability. 2013. Available online: http://pre-sustainability.com (accessed on 9 November 2013).

65. Jackson, W.D.; Christiansen, G. *International Strategic Minerals Inventory Summary Report—Rare-Earth Oxides*; U.S. Geological Survey: Denver, CO, USA, 1993; pp. 1–68.

66. Sprecher, B.; Xiao, Y.; Walton, A.; Speight, J.; Harris, R.; Kleijn, R.; Visser, G.; Kramer, G.J. Life cycle inventory of the production of rare earths and the subsequent production of NdFeB rare earth permanent magnets. *Environ. Sci. Technol.* **2014**, *48*, 3951–3958.

67. Weng, Z.H.; Jowitt, S.M.; Mudd, G.M.; Haque, N. Assessing rare earth element mineral deposit types and links to environmental impacts. *Appl. Earth Sci. Trans. Inst. Min. Metall. B* **2013**, *122*, 83–96.

68. Norgate, T.; Jahanshahi, S.; Rankin, W.J. Assessing the environmental impact of metal production processes. *J. Clean. Prod.* **2007**, *15*, 838–848.

69. Norgate, T.; Lovel, R. *Sustainable Water Use in Minerals and Metal Production*; AusIMM Water in Mining: Brisbane, Australia, 2006; pp. 331–339.

2

CHAPTER

Geological, Mineralogical and Geochemical Aspects for Critical and Rare Metals in Greece

Vasilios Melfos [1,*] and Panagiotis Ch. Voudouris [2]

[1]Department of Mineralogy, Petrology and Economic Geology, Aristotle University of Thessaloniki, Thessaloniki 54124, Greece

[2]Department of Mineralogy-Petrology, University of Athens, Athens 15784, Greece;

ABSTRACT

The European Union (EU) is highly dependent on critical and rare metals which are very important for a sustainable development. However, European industry is not able to cover its demands from native sources and it imports commodities from third countries. Greece is one of the EU countries with the most potential for supplying these strategic metallic raw materials in the future, since it hosts a large number of ore deposits. The epithermal- and porphyry-type deposits and the reduced intrusion related systems of the Serbomacedonian and the Rhodope metallogenic provinces in Northeastern Greece are promising targets for a future exploitation and exploration in Sb, Te, Mo, Re, Ga, In, REE and PGE. Greece is the leading producer of Ni and Al in the EU from laterites and bauxites of central and northern Greece. These deposits also contain significant amounts of Co or REE which should be considered in the future plans of the processing industries. REE are found in high contents at the placer deposits between Chalkidiki and Kavala (North Greece) and elevated PGE concentrations are associated with the chromitites of northwestern Greece. Therefore, the mineral wealth of Greece can contribute significantly to a sustainable and a competitive economy of Europe.

Keywords: critical metals; rare metals; rare earth elements; sustainable development; Greece

1. INTRODUCTION

Raw materials, particularly metals, are very important for sustainable functioning, since they are essential for the high-tech products such as electrical and electronic devices, photovoltaic systems, batteries, catalysts, optical fiber cables, synthetic fuels, *etc*. European industry needs sufficient access to certain mineral raw materials for the sound functioning of the EU's economy. However, the European Union (EU) is highly dependent on imports from third countries of strategically important metals, which are considered to be critical because of their economic value, the high supply risks affected by market distortions and the continuously decrease of their availability. An expert group under the auspices of the European Commission published in 2010, a list of 14 metal raw materials which are considered to be "critical" for the future sustainable development of EU: antimony, beryllium, cobalt, fluorine, gallium, germanium, graphite, indium, magnesium, niobium, platinum group metals, rare earths, tantalum and tungsten [1]. Due to their low concentrations in earth's crust, the critical metals are produced as by-products of other metals which are mined.

During the last years issues concerning the EU policy on securing raw material supplies by improving resource efficiency and recycling, have been high on the political agenda. In November 2008 the new European policy entitled: "The Initiative for raw materials to meet critical needs for growth and employment in Europe" was adopted and promulgated by the EU [2]. This initiative refers to the need of the EU to follow a comprehensive policy for ensuring access to mineral raw materials in order to enhance competitiveness and to achieve the Lisbon partnership for growth and employment.

The EU has many ore mineral deposits, even if their exploration and extraction are facing increased competition for different land uses and a highly protected environment, as well as several limitations in access to these deposits. Greece is one of the most significant metallogenic provinces of EU hosting a large number of ore deposits, which may provide an essential mineral wealth, including critical metals.

Mining has been part of the Greek civilization since prehistoric times, when Greeks began extracting non-metallic minerals and metals for use. Ancient underground mines and ruins of metallurgical furnaces throughout Greece are witnesses of the intense exploitation of the metallic mineral resources, mainly gold, silver, copper, lead and iron. Especially during the Classical and Hellenistic times, famous mining centers operated in Lavrion (Attica), Sifnos and Milos Islands (Aegean sea), Pangeon Mt, Palea Kavala, Chalkidiki and Thasos Island (northern Greece) (Figure 1).

In the last 70 years significant mining operations in Greece have focused on nickel, aluminum, chromium and iron-manganese deposits, as well as lead-zinc-(±Cu, ±Au, ±Ag) sulfide occurrences. However, several of these operations have been stopped in the last years due to various reasons. Recent investigations, especially in northern Greece (Macedonia and western Thrace), have revealed new prospects of gold-silver, copper, molybdenum-rhenium, antimony, tungsten and tellurium ore mineralization. According to [3] the total value of the indicated metallic mineral reserves of Greece exceeds €72 billion. It should be mentioned that Greece is the only country in the EU with large deposits of manganese, chromite, aluminum and nickel. It is the leading producing country in the EU for aluminum (from bauxite) and nickel (from laterite). Significant production of magnesium (from magnesite) is also reported. Greece has a comparative advantage in relation to other European countries due to the presence of a significant mineral wealth.

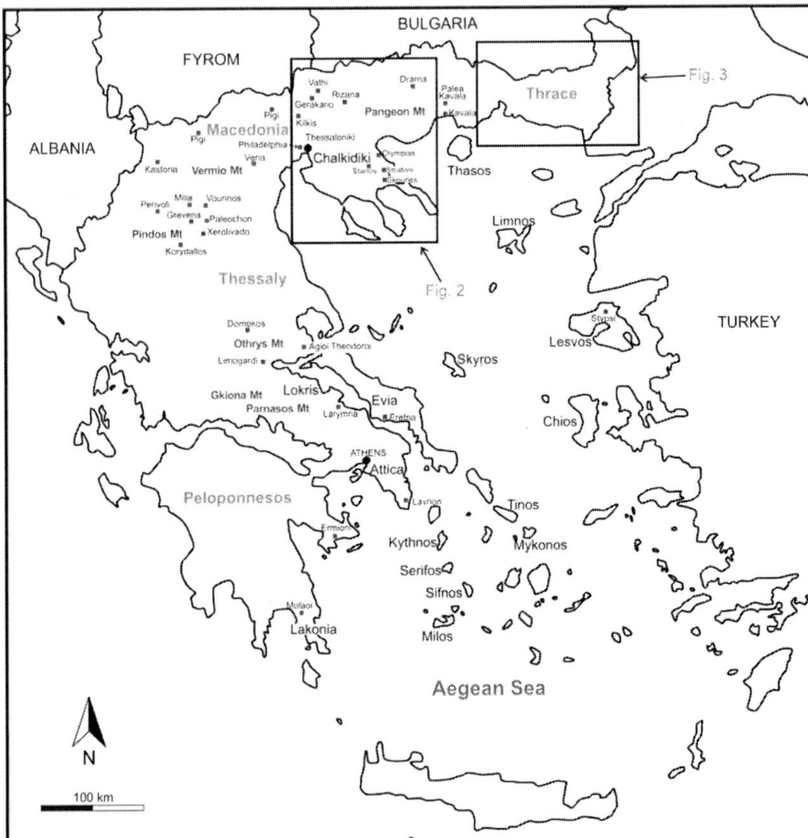

Figure 1. Map of Greece with the distribution of the most significant ore deposits. The ore deposits of Thrace are displayed in Figure 3.

In the last decades, the metal mining industry of Greece operated far below its productive capacity, with an exception of bauxite and laterite. This was affected also by the relatively reduced industrial activities, the long distance from the central European markets which restricted the export potential of the country, and the lack of considerable investments.

In this contribution we present briefly some new data, mainly on Te, In, La and Ce obtained from selected mineralizations (Pagoni Rachi, Perama Hill, Pangeon Mt., Pefka, Vathi/Kilkis) using ICP-MS bulk rock analyses in ACME Laboratories, of Canada. These data are combined with published information on the geology, mineralogy and geochemistry of several critical and rare metallic elements found in Greece, being necessary for the future sustainable development of the EU. We focus mainly on the less investigated metals Sb, Co, Te, Re, Mo, Ga, Ge, In, REE and Platinum Group Elements, because for other commodities like Mn, Cr, Ni, Al and Mg, there is a broad knowledge and experience with mining in Greece. These data will contribute significantly to the knowledge about the presence of critical metals in specific regions of Greece, for a possible future exploration and exploitation.

2. GEOLOGICAL OVERVIEW AND VARIOUS ORE DEPOSIT STYLES IN GREECE

Geological setting of Greece consists of various elongated geotectonic terrains which represent successive episodes of oceanic riftings, spreadings, subductions and collisions in the region of Neotethys, known as Hellenide orogen [4]. The evolution of the Neotethys Ocean and the related lithosphere plates caused magmatic processes which led to the formation of considerable ore mineralizations.

Large magmatic Cr-(±PGE) deposits (Vourinos, Xerolivado, Milia, Korydallos/Pindos, Eretria, Domokos/Othrys, Chalkidiki and eastern Thessaly) (Table 1) are related with Mesozoic ophiolitic chains along the Pindos and Vardar ocean sutured zones. Ni-Fe laterites (Lokris, Evia, Kastoria) and Al bauxites (Parnassos Mt, Gkiona Mt) are also associated with these ophiolites, although most of these deposits are allochthonous, being incorporated among various karstified limestones due to the subsequent sedimentary processes. Only few, mostly uneconomic, VMS-type mineralizations (Table 1) are also in close association with the Mesozoic ophiolitic sequences within meta-volcanosedimentary rocks of basic affinity. They are mainly of Cu-pyrite Cyprus-, Kuroko Zn-Pb-Ag- and Fe-Cu-Ni-Co-type, forming stratabound and stratiform ore bodies in Xylagani (Thrace), in Vermio (Veria), in Perivoli (Pindos), in Agioi Theodoroi, Limogardi (Othrys), Eretria (Evia) and in Ermioni and Molaoi/Lakonia (Peloponnesos) (Table 1, Figure 1).

Significant precious and base metal sulfide ore deposits are widespread in Attic-Cycladic ore belt (south Evia, Sifnos, Mykonos, Tinos, Kythnos), along the South Aegean Active Volcanic Arc (Milos) and in Attica (Lavrion), of Lower Pliocene to Miocene age [5]. However, the most promising regions for future critical- and rare-metal exploration projects are the Serbomacedonian and Rhodope metallogenic provinces (Figure 2 and Figure 3), in northeastern Greece [6]. These provinces include skarn-(Kimmeria/Xanthi), carbonate-replacement-(Olympias, Stratoni/NE Chalkidki), reduced intrusion-related-(Palea Kavala), epithermal (Perama Hill, Mavrokoryfi, Kassiteres, Sapes, Agios Philippos/Kirki, Pefka/Evros) and porphyry-(Pagoni Rachi/Kirki, Kassiteres-Sapes, Maronia, Skouries/Chalkidiki, Vathi and Gerakario/Kilkis) deposit styles, as well as placer gold occurrences and supergene fault/karst-controlled manganese oxides (Drama, NE Chalkidiki) and gold-bearing iron oxides (Palea Kavala, Thasos) [6,7]. These deposits are characterized by Pb, Zn, Sb, Bi, Cu and Mo ores with, less commonly, economic concentrations of Fe and Sn. They also include minor Au, Ag, PGE, Te, U, W, Re and As occurrences [7].

The Serbomacedonian and Rhodope metamorphic massifs comprise the crystalline basement of the Alpine orogenic belt in the Balkan peninsula and consist of amphibolite facies metamorphic rocks of continental and oceanic origin (Figure 2 andFigure 3). They demonstrate a structural record of a late Mesozoic deformation episode overprinted by an early Cenozoic extensional deformation [8]. The deposits of the Serbomacedonian and Rhodope metallogenic provinces were formed during the final stage of the Tertiary orogenic collapse, which led to the formation of widespread Oligocene-Miocene silicic to intermediate magmatism [7,9,10,11]. This magmatism, considered to be part of the Cretaceous to Miocene Alpine-Balkan-Carpathian-Dinaride orogen [11], resulted from subduction and collision of the Pelagonian and Apulian microcontinents, detached from Africa in Triassic, with the European plate [12]. Slab detachment and/or roll-back was the principal mechanism, which resulted in the local ascent of asthenospheric mantle into the metasomatized mantle lithosphere and to magma generation [9,13]. Magmas were intruded to shallow depths along deep seated detachment faults,

Table 1. Characteristics of critical- and rare metal-bearing mineralization in Greece (data sources in text).

Mineralization-type	Province	Area	Main metals	Critical and rare metals
Magmatic Cr-(±PGE)	Pindos	Vourinos	Cr	PGE
	Pindos	Xerolivado	Cr	PGE
	Pindos	Milia	Cr	PGE (Os, Ir, Ru)
	Pindos	Korydallos	Cr	PGE (Pt, Pd, Ru, Rh)
	Evia	Eretria	Cr	–
	Othrys	Domokos	Cr	–
	Thessaloniki	Triadi-Vassilika	Cr	PGE
	Chalkidiki	Vavdos	Cr	PGE
	Chalkidiki	Gerakini-Ormylia	Cr	PGE
	Eastern Thessaly	Kallipefki, Agia	Cr	–
	Aegean sea	Skyros	Cr	PGE (Ru)
	Veria	Vermio	Cr	PGE (Os, Ir, Ru, Rh, Pt, Pd)
VMS (Cyprus- and Kuroko-type)	Thrace	Xylagani	Cu, Fe, Au	–
	Serres	Lagadi	Cu, Fe	Co
	Veria	Vermio	Fe, Cu, Ni, Co	Co
	Kilkis	Polykastro	Pb, Zn, Ag	In, Ge, Ga
	Pindos	Perivoli	Fe, Cu	Co
	Othrys	Agioi Theodoroi	Fe, Cu, Ni, Co	Co
	Othrys	Limogardi	Fe, Cu, Ni, Co	Co
	Peloponnesos	Ermioni	Cu, Fe	Co
	Peloponnesos	Molaoi/Lakonia	Zn, Pb, Ag	In, Ge, Ga
Ni-Fe laterites	Lokris	Larymna	Ni, Fe	Co
	Evia		Ni, Fe	Co
	Kastoria	Ieropigi	Ni, Fe	Co
	Grevena	Paleochori	Ni, Fe	Co
	Veria	Vermio	Ni, Fe	Co
Al bauxites	Parnassos		Al	REE
	Gkiona		Al	REE
Shear-zone related	Chalkidiki	Stanos	Cu, Au, Bi, Mo	Te
	Lachanas	Rizana	Sb, W	Sb
	Thessaloniki	Philadelphia	Sb, Fe, Au	Sb
Carbonate replacement	Chalkidiki	Olympias	Pb, Zn, Au, Ag	In, Ge, Ga
	Chalkidiki	Stratoni	Pb, Zn, Ag	Te
	Xanthi	Thermes	Pb, Zn, Ag	In, Ge, Ga
	Attica	Kamariza/Lavrion	Pb, Zn, Ag, Au	In
Reduced intrusion-hosted	Kilkis	Pigi	Mo	W
	Thrace	Kimmeria/Xanthi	Fe, Cu, Mo, Bi	W
	Kavala	Palea Kavala	Au, Bi	Te
	Kavala-Drama	Pangeon Mt	Au, Bi	Te
	Attica	Plaka/Lavrion	Cu, Pb, Zn	Ga, REE (La, Ce)

Mineralization-type	Province	Area	Main metals	Critical and rare metals
Epithermal	Thrace	Perama Hill	Cu, As, Au, Bi	Te
		Mavrokoryfi	Cu, Sb, Ag, Au	Te
		St. Demetrios/Sapes	Cu, As, Au, Ag	Te
		St. Barbara/ Kassiteres	Cu, Au, Ag	Te
		Agios Philippos/Kirki	Pb, As, Cu, Ag, Bi	In, Ga, Ge
		Pefka/Evros	Cu, As, Ag, Au	Te, In,
		Kalotycho	Fe, Cu	–
		Kallyntirion	Pb, Zn, Cu, Au, Ag	Sb
	Aegean Sea	Tinos Island	Cu, Pb, Au, Ag	Te
		Limnos Island	Pb, Zn, Cu, Au, Ag	Te
Porphyry	Thrace	Pagoni Rachi/Kirki	Cu, Au, Mo	Te, Re
		Konos/Sapes	Cu, Au, Mo	Te, Re
		Maronia	Cu, Au, Mo	Re
		Myli/Esymi	Cu, Au, Mo	–
		Melitena	Cu, Mo	Re
	Chalkidiki	Skouries	Cu, Au, Mo	PGE (Pd, Pt)
	Kilkis	Vathi	Cu, Au, Mo	REE (La, Ce)
		Gerakario	Cu, Au	Sb
	Attica	Plaka/Lavrion	Mo	W
	Aegean Sea	Sardes/Limnos Island	Mo	–
		Fakos/Limnos Island	Cu, Mo	–
		Stypsi/Lesvos Island	Mo, Bi	–
Supergene fault /karst-controlled Mn-oxides	Drama		Mn	–
	NE Chalkidki		Mn	–
Fe-oxides	Kavala	Palea Kavala	Au	–
	Thasos		Au	–
Sediment-hosted, placers	Veria	Vegora	P	Ga
	Aegean Sea	Chios Island	–	Sb
	Chalkidiki-Kavala	Coastal zone	–	REE (Ce, La, Nd, Pr)

forming mafic to intermediate mantle-derived plutonic, subvolcanic and volcanic rocks of calc-alkaline, high-K calc-alkaline and shoshonitic to ultra-potassic (lamprophyric) affinities.

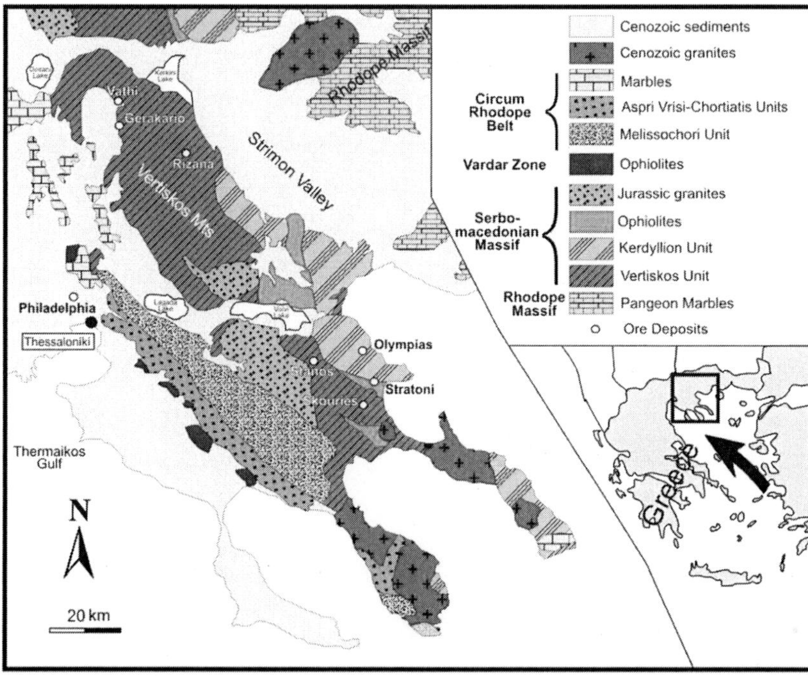

Figure 2. Simplified geological map of the Serbomacedonian metallogenic province in Greece, displaying the major ore deposits and prospects.

3. CRITICAL AND RARE METALS IN GREECE

3.1. Antimony (Sb)

Antimony is used for a number of industrial applications, as diodes and infrared detectors in the semiconductor manufacturing. It is alloyed with lead to increase its hardness. Antimony compounds are used in batteries, low friction metals, paints, ceramics, glass and pottery, as flame retardants in adhesives, plastics, rubber and textiles. Stibnite (Sb_2S_3) is the predominant ore mineral of antimony. China supplies over 75% of the total world production and in 2010 introduced quantitative and export licensing restrictions. Therefore antimony was recently placed in the highest risk category for future supply disruption.

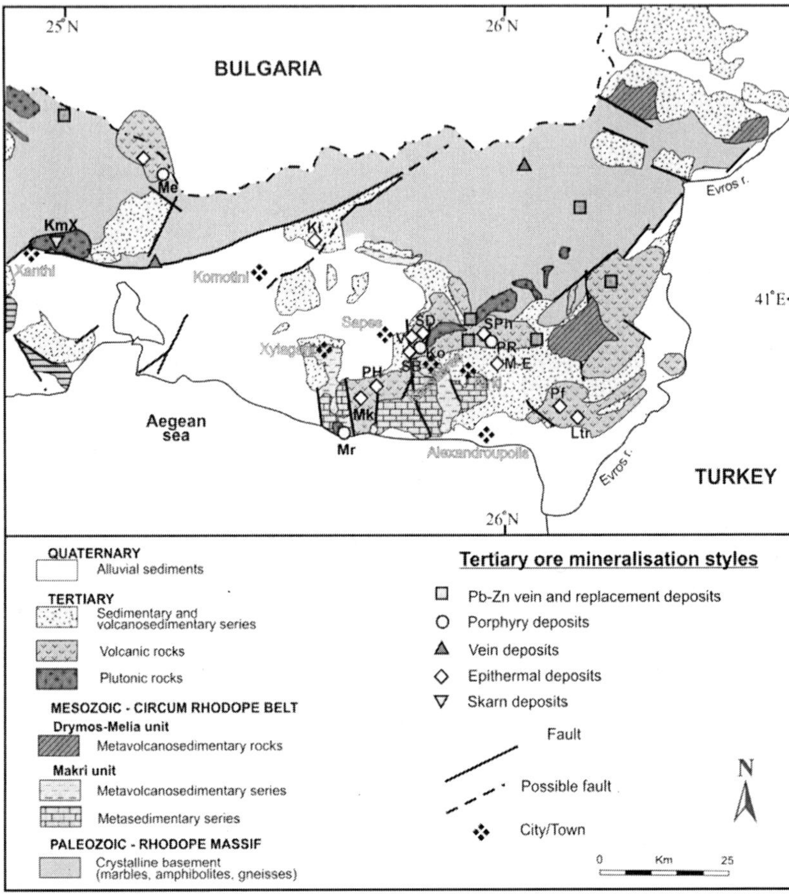

Figure 3. Simplified geological map of the Rhodope metallogenic province in Greece displaying the major Tertiary ore deposits and prospects [5,6 with modifications]. KmX: Kimmeria/Xanthi; Me: Melitena; Mr: Maronia; Ko: Konos/Sapes; SB: St. Barbara/Kassiteres; SD: St. Demetrios/Sapes; V: Viper; PH: Perama Hill; Mk: Mavrokoryfi; PR: Pagoni Rachi/Kirki; SPh: Agios Philippos/Kirki; Pf: Pefka; Ltr: Loutros; Kl: Kallyntirion; M-E: Myli/Esymi.

In Greece, antimony is found in quartz veins mainly in the form of stibnite. One of the most important Sb occurrences is the low-sulfidation epithermal system of Kallyntirion prospect at the Rhodope metallogenic province (Figure 3), also including Pb, Zn,

Cu, Au and Ag. Another significant stibnite mineralization is the shear zone-related Rizana/Lachanas deposit, Serbomacedonian province, which is related to sheeted quartz veins crosscutting gneisses and amphibolites. The mineralization is spread over an area 10 km long and 2–4 km wide, and is potential for future exploitation. In the years 1931–1938 mining operations produced 2.3×10^{-6} kg of ore consisting more than 30% Sb [14]. The estimated reserves are over 5.0×10^{-7} kg with 30%–35% Sb, in addition to small tungsten concentrations, associated with wolframite [14]. According to [15] the chemical analyses of enriched samples from Rizana/Lachanas revealed Sb content between 71.73 and 83.87 wt % (average 75.77 wt %). Other stibnite occurrences with relatively low Sb contents are located in Philadelphia/Thessaloniki, Chios Island and Gerakario/Kilkis [14].

Except from stibnite, antimony is also incorporated in numerous sulfosalts (tetrahedrite, famatinite, chalcostibite, zinkenite, bournonite, boulangerite) which are common in porphyry-, epithermal- and intrusion-related systems at the Rhodope metallogenic province, such as in Maronia, Pagoni Rachi/Kirki, Agios Philippos/Kirki, Mavrokoryfi, Pefka, Perama Hill, Sapes, Kapsalina/Thasos, Pangeon Mt. In some cases mineralized samples contain over 0.2 wt % Sb [6,7,16,17]. In addition, the galena concentrates from Lavrion, Attica, contain from 0.3 to 0.7 wt % Sb [18].

3.2. Cobalt (Co)

Cobalt has been used for a long time as a pigment in glass and porcelain industry. Recently it was classified as one of the critical metals due to its unique applications in lithium ion batteries, in resistant superalloys, in green-energy systems and in magnets [19]. Globally, cobalt is mined as a by-product of copper and nickel deposits in the Republic of the Congo-DRC (40%), in Zambia (20%), in Canada, in Australia, in Cuba, *etc.* China imports large amounts of cobalt (almost all the DRC's production) in order to produce refined cobalt which is exported to the USA and the EU.

Greece is the major nickel supplier in EU and produces 2%–3% of the world total Ni. Nickel is mined from the laterite deposits in Lokris, Evia and Kastoria and is recovered by pyrometallurgical methods. These Fe-Ni laterites, which also contain cobalt, are mainly allochthonous and are associated with Mesozoic ophiolites. They are intercalated in Mesozoic limestones and show multistage transportation and re-deposition, responsible for a post-sedimentary redistribution of metals, resulting in Mn, Co, Ni-rich zones [20,21]. Chemical composition of different laterite occurrences in Greece shows that cobalt is mainly concentrated in the Fe-Ni-rich zones, demonstrating elevated contents which reach 0.10% in Paleochori, Grevena, 0.14% in Vermio, 0.16% in Kastoria and 0.22% in Lokris [20,21,22]. Co in these deposits is concentrated mainly in the mineral asbolane [22]. The metallurgical process of the Greek laterites in Larymna by LARCO, which produces Fe-Ni alloy, does not allow any Co recovering, which can be separated by hydrometallurgical treatment [23]. Consequently, Greece remains a potential region to supply the EU with significant amounts of cobalt in the future.

It should be also mentioned here that the non-economic sulfide ores hosted in the ophiolite complexes in Greece, including Cyprus-type and Fe-Cu-Ni-Co type, at Pindos and Othrys contain up to 2300 ppm Co [24].

3.3. Tellurium (Te)

Tellurium, a relatively rare metalloid, is used in manufacturing alloys and semiconductors, as well in vulcanizing rubber and in catalysts for petroleum. In alloys it improves properties like ductility and tensile strength or increase the durability and flexibility. It is added to lead to decrease the corrosive action of sulfuric acid in batteries and to improve the lead's strength and hardness. It is also used in making various types of detectors. Compounds made of Te, Cd and Hg, are excellent infrared sensitive semiconductors [25]. The use of high-purity tellurium enhances the effectiveness of the solar cell electric power generations and is very promising for future applications. Tellurium is extensively used as a coloring agent in ceramics and glasses. It is also used as a gasoline additive to reduce engine knocks in automobiles.

Almost all tellurium currently produced is obtained as a by-product of copper refining. It is estimated that the Cu deposits globally contain approximately 2.2×10^{-7} kg of tellurium reserves worldwide [25]. However, the high-grade ores are being exhausted and this may result in limitations in future tellurium supply.

In Greece tellurium is a common constituent of Cu-Au epithermal high-intermediate sulfidation type mineralizations (St. Barbara/Kassiteres, St. Demetrios/Sapes, Perama Hill, Mavrokoryfi, Pefka) and spatially associated porphyry Cu-Mo systems (Pagoni Rachi/Kirki, Konos/Sapes), in intrusion-related Au-Bi-Te systems (Palea Kavala, Pangeon Mt) of Oligocene-Miocene age at the Rhodope province, as well as in Limnos and Tinos Islands, Aegean sea (Fakos porphyry-epithermal- and Panormos intrusion-related Cu-Au-Te prospects respectively) [7,26,27,28]. Tellurium in these prospects is incorporated in various telluride phases, such as hessite, stützite, petzite, altaite, sylvanite, melonite, calaverite, coloradoite and tetradymite, as well as in the sulfosalt goldfieldite (tetrahedrite-group). In rare cases native tellurium is present in minor concentrations (Pefka, Tinos, Palea Kavala). The tellurium content in the mineralized samples is variable and is related with the distribution of Te-bearing minerals. According to new chemical analyses, provided here for the first time, the richest samples contain up to 40 ppm in Pagoni Rachi, 45 ppm in Perama Hill, 60 ppm in Pangeon Mt, and >1000 ppm in Pefka. Comparing with other potential Te-rich ore deposits worldwide (Boliden mining company will produce Te from the Kankberg gold mine, Sweden, with reserves of 2.8 Mt which contains 186 ppm Te in average [29]), the Pefka area is the most promising prospect for future Te exploration and exploitation in Greece.

The source rocks of the tellurides in northeastern Greek mineralizations are mantle-derived calc-alkaline (Thrace) to shoshonitic (Limnos) subvolcanic intrusions. Magmas were likely enriched in Te as a result of previous subduction of Te-bearing sediments [7].

3.4. Rhenium (Re) and Molybdenum (Mo)

Rhenium is very resistant against corrosion and is used in high-temperature superalloys for jet engines and industrial gas turbine engines. It improves the properties of tungsten and alloys in X-ray sources and in thermocouples to measure temperatures up to

2200 °C. Rhenium is also used in filaments, in chemical catalysts at petroleum industry, in space craft thrusters, in electrical contacts, in coatings, in flash lamps for photography. Radioactive rhenium is used in prevention and treatment of restenosis and liver cancer.

Molybdenum has various applications in the construction of grade and stainless steel, in chemicals such as lubricants, in the manufacturing of tool and high-speed steels, in the manufacture industry of cast iron and molybdenum metals, as well as in super alloys.

The world's highest grade molybdenum-rhenium deposit is located in Merlin at the Mount Dore ore prospect, consisting mainly of molybdenite, with minor chalcopyrite and pyrite, and is hosted within a series of altered carbonaceous metapelites and phyllites [30]. The mineralization exceeds over 1 km in length and is up to 40 m thick, to approximately 550 m from the surface. The current resources average 0.6% Mo and 10 ppm Re.

Molybdenite occurs mainly in three mineralization styles in Greece [31]: (1) Porphyry Cu-Mo and Mo ± Cu (Pagoni Rachi/Kirki, Ktismata/Maronia, Myli/Esymi, Konos/Sapes, Skouries/Chalkidiki, Melitena, Sardes/Limnos, Fakos/Limnos, Stypsi/Lesvos); (2) Reduced intrusion-hosted Mo-W systems (Kimmeria/Xanthi, Pigi/Kilkis, PlakaP/Lavrion, Serifos Island); (3) Shear zone related Cu-Au-Bi-Mo (Stanos/Chalkidiki).

A small molybdenite production was operated in Pigi in 1940–1944 with variable Mo grade, from 0.3% to 1.3% (rarely up to 10%) with an average of 0.6%–0.7% [14]. There is not any estimation about the current Mo reserves.

The porphyry Cu-Mo and Mo ± -Cu prospects of Pagoni Rachi, Maronia and Melitena, in western Thrace contain molybdenite with high to ultrahigh Re content (reaching 4.7 wt %), which are the highest concentrations of Re-in-molybdenites ever measured [31,32,33]. The microprobe analyses demonstrate that there is a positive correlation between Mo and Re supporting the concept that Re substitutes for Mo in the structure of molybdenite. Moreover, the very rare mineral rheniite (ReS_2) is found in the porphyry systems of the Pagoni Rachi and Konos/Sapes prospects [31,33].

Rheniite and high-Re molybdenite in these deposits precipitated under oxidizing conditions and from relatively acid hydrothermal solutions. The extremely high contents of rhenium in molybdenite at the Greek porphyry prospects could be explained by a rhenium release from the asthenospheric mantle wedge by subduction-related fluids and melts [33]. The Re-enrichment in the western Thrace molybdenites, suggest that this region should be a potential target for future Re- and Mo exploration and exploitation.

3.5. Gallium (Ga), Germanium (Ge) and Indium (In)

The rare metals gallium, germanium and indium are the most important commodities in the high-technology society, essential for electronic components. They are critical metals for the efficiency of utilization of recourses that are used and consumed in the order of millions of tones [34]. Moreover, germanium has major applications in optical materials and indium in liquid crystal displays (LCD). Presently gallium is produced as a by-product of bauxite processing and from the zinc processing residues, although the largest reserves are associated with phosphate ores [19]. Germanium and indium are recovered during the zinc production process.

The epithermal polymetallic veins of the Rhodope province are the most promising ore types for Ga, Ge and In exploitation in Greece. The Agios Philippos mine at Kirki, western Thrace, is a polymetallic vein deposit with intermediate-high sulfidation epithermal affinities and contains an unusual mineralogy consisting of several Pb-As-Cu-Bi-Sn sulfosalts. It is hosted within Eocene sediments and displays spatial relationships to Upper Oligocene rhyolite porphyry exposed at the northern part of the mining area. Wurtzite is the dominant Zn-ore mineral in Agios Philippos, along with sphalerite, and is found in the form of veinlets and granular masses. It contains considerable amounts of In (reaching 3.5 wt %) and Ga (reaching 1.6 wt %), whereas Ge is present in smaller amount (below 0.3 wt %) [35].

In Pefka/Evros the high-intermediate sulfidation epithermal mineralization is related to quartz veins hosted within andesitic and dacitic volcanic rocks. Bulk ore analyses (present study) revealed relatively high In content (up to 675 ppm), whereas Ga and Ge are low (max 17 ppm and 16 ppm respectively). Another potential deposit for future Ga exploration is the Lavrion intrusion-related deposit, especially the Pb-Zn-Cu bearing granitic dikes from Plaka deposit, where Ga reaches 326 ppm [36].

Relatively lower contents of In are indicated for the Lavrion carbonate-replacement ore bodies where sphalerite concentrates contain up to 43 ppm and the bulk Cu-pyritic ores up to 72 ppm [18]. The Polykastro/Kilkis and Molaoi/Peloponessos (Kuroko-type) and the Thermes/Xanthi and Olympias/Chalkidiki (carbonate-replacement) deposits should be also evaluated for their In, Ge and Ga potential [3].

Phosphate ores represent the largest reserves of several million tons, generally with a low Ga-content of 0.01%–0.1% [19]. The Neogene phosphate deposits of Greece contain low Ga content in general. The highest concentrations have been referred in the Vegora lacustrine basins of Vegora (Vegora-Komnina), 40 km northwest from Veria, where Ga in the phosphatic rocks reaches 15.9 ppm [37], and they cannot be considered as potential sources for future exploitation.

3.6. Rare Earth Elements (REE)

The demand for rare earth elements (REE) has been increased in the last years due to their applications in a wide range of consumer products such as automotive catalytic converters, photovoltaic systems, wind turbines, flat panel displays (cell phones, screen TVs, computers, portable DVDs), permanent magnets, hybrid and electric vehicles. They are also strategic raw materials with important defense uses in jet fighter engines, space-based satellites and communication systems. China controls approximately 95% of the global supply of rare earths and recently imposed quotas on exports in order to preserve its resources for domestic consumption. This has caused concerns in the EU, USA and Japanese markets regarding future REE supplies.

Recently it was announced that by 2013 USA will be able to meet a majority of its demands from North American sources [38]. European industry imports more than 90% of its REE needs and consumes about 30% of the global annual production [38]. The abandoned mines in Greece could relieve European industry problems [39] and this is in accordance with the recent investigation results in several areas of Greece.

Significant reserves of REE are found in the placer deposits along the coast of Northeastern Greece between the peninsula of Chalkidiki and Kavala, with 240 km coastal length [40,41,42]. Geochemical investigations revealed up to 8000 ppm in REE with 93.5% consisting of the light REE mainly Ce, La, Nd and Pr, with allanite being the main REE-bearing mineral phase [40]. It is very probable that the primary sources of these LREE bearing placers are located at the nearby Kavala or Pangeon Mt. intrusion-related systems. Recently, one metallurgical scoria from old smelting furnaces (possibly from Ottoman period) at the Pangeon Mt., revealed 6171 ppm La, 10,883 ppm Ce, 1062 ppm Pr, 3249 ppm Nd, 334 ppm Sm and 214 ppm Eu [43]. Further geological and geochemical research could help locate the extremely high LREE-enriched source rocks, possibly on Pangeon Mt. It should be however mentioned that the heavy REE are the more highly sought after in the world industry.

Of major importance for future REE exploration-exploitation can also be considered the bauxite deposits of Parnasos Mt, with REE contents reaching 6300 ppm [3], the Pb-Zn-Cu bearing granitic dikes from Plaka deposit in Lavrion with 428 ppm La and 976 ppm Ce [36], and the porphyry Cu-Au system of Vathi/Kilkis with relatively high content of La (500 ppm) and Ce (715 ppm) [this study].

3.7. Platinum Group Elements (PGE)

An increased worldwide usage of platinum group elements (PGE = Pt, Os, Ir, Pd, Ru, Rh) has been observed during the last decades, due to their unique physical and chemical properties that make them critical to many emerging technologies. Consequently they have become essential commodities and are considered as strategic metals because of their specialized applications in chemical, petroleum, high-tech, electrical, medical, and aerospace industries [19]. They also have important uses in environmentally related technologies, such as automobile catalytic converters and fuel cells. The global reserves of PGE are mainly concentrated in

South Africa, followed by Russia and Canada, where they are by-products of nickel processing [19].

In Greece there are two target mineralization groups that contain elevated PGE grades: Porphyry-type deposits and chromite ores. A significant Pd-Pt enrichment has been determined in the Skouries porphyry Cu-Au deposit of NE Chalkidiki peninsula, in the copper ore, as well as and in the flotation concentrates [44,45]. The Pd content in the mineralized samples (stockwork veinlets from the potassic alteration and from the oxidation zones) ranges from 45 to 490 ppb and up to 3300 ppb in chalcopyrite concentrates, suggesting that Pd (in the form of the telluride merenskyite) is mainly associated with chalcopyrite [44]. The leaching process of the final concentrate (21 wt % Cu, 22 ppm Au, 2.4 ppm Pd, 0.04 ppm Pt) resulted in the recovery of 96.6% Au, 97.7% Pd and 100% Pt [45]. Since the intended investment in the Skouries deposit focuses on gold and copper mining and metallurgy, the recovery of Pd and Pt has a positive economic potential for a future sustainable development.

The second target mineralization group for PGE exploration in Greece is associated with the ophiolites of Veria, Pindos and Skyros Island. High PGE concentrations, reaching 25 ppm total PGE's (Os = 7400, Ir = 6020, Ru = 9700, Rh = 310, Pt = 760, Pd = 750, all in ppb) were determined in chromitites hosted in the Veria ophiolites, northern Greece [46,47]. These elevated PGE contents are associated with a great number (>100) of different PGM grains, mainly Os-Ru-Ir-Fe alloys, reaching 1 mm in size [47]. They are concentrated along a shear zone of strongly brecciated chromite ore. In Korydallos area, Pindos the PGE-mineralization is associated with several PGM grains (Pd-Cu alloys, sperrylite, Pt-Fe-Ni alloys, naldrettite, bismuthotelluride, native Au, *etc.*) [48]. The total PGE + Au content in the chromitites with low Cr# chromian spinel is 28 ppm (Pt = 17100, Pd = 7860, Ru = 2100, Rh = 1140, all in ppb) among the highest ever measured in ophiolitic chromitites worldwide, and the highest ever mentioned for chromitites from Greece [48]. It is assumed that a sulfur bearing melt played an important role in scavenging the PGE + Au content of the silicate magma from which chromian spinel had already started to crystallize [48].

The ophiolitic complexes of Thessaloniki (Triadi, Vassilika) and Chalkidiki (Vavdos, Gerakini-Ormylia) contain relatively low PGE contents (<0.5 ppm) and so they are not of economic significance [49].

4. CONCLUSIONS

The Serbomacedonian and Rhodope metallogenic districts in Greece (Figure 2 and Figure 3) are among the richest ore provinces in Europe, and can be an exploration target in the near future, especially for critical and rare metals. The epithermal- and porphyry-type mineralizations, and the reduced intrusion-related systems of Northeastern Greece could produce Sb, Te, Mo, Re, Ga, In and PGE as by-products, along with base metals, Au and Ag, if the mining and extractive activities will start. In addition, laterites and bauxites of central and northern Greece which are mined for Ni and Al respectively, contain significant amounts of Co or REE and the processing industries should focus on the probability for a future exploitation. The chromitites of northwestern Greece contain significant concentrations of PGE, whereas the coastal sediments between Chalkidiki and Kavala are characterized by REE geochemical anomalies which may be related with the Kavala and Pangeon Mt reduced intrusion-related system, although this has to be clarified.

The mineral wealth of Greece can provide solutions to the challenge of a sustainable and a competitive economy. The rational exploitation of the ore deposits will offer significant developmental benefits and it will enable Greece to strengthen its global role in the mining industry. A new mining policy must focus on the more effective exploitation of the mineral resources, the improvement of the environmental performance, the safe operation of the mines and the accidents prevention, the mining waste management and the recycling process. However, when considering the exploitation of critical and rare metals, we have to keep in mind that this process is long, entails a high risk and requires large investments.

ACKNOWLEDGMENTS

Special thanks are due to Basilios Tsikouras and two anonymous reviewers for carefully reviewing the manuscript and for their constructive comments and suggestions, resulting in significant improvement of this paper.

REFERENCES

1. Commission of the European Communities. *Critical Raw Materials for the EU:Report of the ad-hoc Working Group on Defining Critical Raw Materials*, Available online: http://ec.europa.eu/enterprise/policies/raw-materials/documents/index_en.htm (accessed on 16 October 2012).
2. Commission of the European Communities. *Commission Staff Working Document*, Available online: http://ec.europa.eu/enterprise/sectors/metals-minerals/files/sec_2741_en.pdf (accessed on 16 October 2012).
3. Tsirambides, A.; Filippidis, A. Exploration key to growing Greek industry. *Ind. Miner.* **2012**, 44–47.
4. Pe-Piper, G.; Piper, D.J. *The Igneous Rocks of Greece*; Gebroder Borntraeger: Berlin, Germany, 2002.
5. Skarpelis, N. Geodynamics and evolution of the miocene mineralization in the Cycladic-Pelagonian belt, Hellenides.*Bull. Geol. Soc. Greece* **2002**, *34*, 2191–2206.
6. Melfos, V.; Vavelidis, M.; Christofides, G.; Seidel, E. Origin and evolution of the tertiary Maronia porphyry copper-molybdenum deposit, Thrace, Greece. *Miner. Deposita* **2002**, *37*, 648–668.
7. Voudouris, P. Comparative mineralogical study of tertiary Te-rich epithermal and porphyry systems in northeastern Greece. *Mineral. Petrol.* **2006**, *87*, 241–275.

8. Bonev, N.; Dilek, Y.; Hanchar, J.M.; Bogdanov, K.; Klain, L. Nd-Sr-Pb isotopic composition and mantle sources of triassic rift units in the Serbo-Macedonian and the western Rhodope massifs (Bulgaria–Greece). *Geol. Mag.* **2012**, *149*, 146–152.

9. De Boorder, H.; Spakman, W.; White, S.H.; Wortel, M.J.R. Late cenozoic mineralization, orogenic collapse and slab detachment in the European Alpine belt. *Earth Planet. Sci. Lett.* **1998**, *164*, 569–575.

10. Marchev, P.; Kaiser-Rohrmeier, M.; Heinrich, Ch.; Ovtcharova, M.; von Quadt, A.; Raicheva, R. Hydrothermal ore deposits related to post-orogenic extentional magmatism and core complex formation: The Rhodope Massif of Bulgaria and Greece. *Ore Geol. Rev.* **2005**, *27*, 53–89.

11. Heinrich, C.A.; Neubauer, F. Cu-Au-Pb-Zn-Ag metallogeny of the Alpine-Balkan-Carpathian-Dinaride geodynamic province. *Miner. Deposita* **2002**, *37*, 533–540.

12. Pe-Piper, G.; Christofides, G.; Eleftheriadis, G. Lead and neodymium isotopic composition of Tertiary igneous rocks of northeastern Greece and their regional significance. *Acta Vulcanol.* **1995**, *10*, 255–263.

13. Blundell, D.; Arndt, N.; Cobblod, P.R.; Heinrich, C. Processes of tectonism, magmatism and mineralization: Lessons from Europe. *Ore Geol. Rev.* **2005**, *27*, 333–349.

14. Maben, B.F.; Zigdis, I.G. *The Mineral Wealth of Greece*; UNRRA: Athens, Greece, 1947.

15. Vasilatos, Ch.; Mparlas, K.; Stamatakis, M.; Tsivilis, S. Wolframite-stibnite mineral assemblages from Rizana Lachanas, Macedonia, Greece, and their possible use as flux agent in the manufacturing of clinker. *Bull. Geol. Soc. Greece* **2001**, *3*, 827–834.

16. Voudouris, P.; Papavasiliou, C.; Melfos, V. Silver mineralogy of St Philippos deposit (NE Greece) and its relationship to a Te-bearing porphyry-Cu-Mo mineralization. *Geoch. Mineral. Petrol.* **2005**, *43*, 155–160.

17. Eliopoulos, D.G.; Kilias, S.P. Marble-hosted submicroscopic gold mineralization at Asimotrypes area, mount Pangeon, southern Rhodope core complex, Greece. *Econ. Geol.* **2001**, *106*, 751–780.

18. Skarpelis, N. The Lavrion deposit (SE Attika, Greece): Geology, mineralogy and minor elements chemistry. *N. Jb. Miner. Abh.* **2007**, *183*, 227–249.

19. Buchert, M.; Schüler, D.; Bleher, D. Critical metals for future sustainable technologies and their recycling potential. Available online: http://resourcefever.com/detail/ items/ critical-metals-for-sustainable-technologies-and-their-recycling.html (accessed on 11 October 2012).

20. Eliopoulos, D.G.; Economou-Eliopoulos, M. Geochemical and mineralogical characteristics of Fe-Ni- and bauxitic-laterite deposits of Greece. *Ore Geol. Rev.* **2000**, *16*, 41–58.

21. Eliopoulos, D.G.; Economou-Eliopoulos, M.; Apostolikas, A.; Golightly, J.P. Geochemical features of nickel-laterite deposits from the Balkan peninsula and Gordes, Turkey: The genetic and environmental significance of arsenic. *Ore Geol. Rev.* **2012**, *48*, 413–427.

22. Apostolikas, A.; Frogoudakis, E.; Maglaras, K. Nickel-bearing deposits in Western Macedonia, Greece: Present and perspectives. In Proceedings of the 1st Conference of the Committee of Economic Geology, Mineralogy and Geochemistry of the Geological Society of Greece, Kozani, Greece, 11–14 May 2000.

23. Herrington, R. Potential for cobalt recovery from lateritic ores in Europe. In *Geophysical Research Abstracts*; European Geosciences Union: Vienna, Austria, 2012.

24. Economou-Eliopoulos, M.; Eliopoulos, D.; Chryssoulis, S. A comparison of high-Au massive sulfide ores hosted in ophiolite complexes of the Balkan Peninsula with modern analogues: Genetic significance. *Ore Geol. Rev.* **2008**, *33*, 81–100.

25. Moss, R.L.; Tzimas, E.; Kara, H.; Willis, P.; Kooroshy, J. Critical metals in strategic energy technologies—assessing rare metals as supply-chain bottlenecks in low-carbon

energy technologies. Available online: http://publications.jrc.ec.europa.eu/repository/handle/111111111/22726 (ac cessed on 11 October 2012).

26. Fornadel, A.P.; Spry, P.G.; Melfos, V.; Vavelidis, M.; Voudouris, P. Is the Palea Kavala Bi-Te-Pb-Sb ± Au district, northeastern Greece, an intrusion-related system? *Ore Geo. Rev.* **2011**, *39*, 119–133.

27. Fornadel, A.P.; Voudouris, P.; Paul, G.S.; Melfos, V. Mineralogical, stable isotope, and fluid inclusion studies of spatially related porphyry Cu and epithermal Au-Te mineralization, Fakos Peninsula, Limnos Island, Greece. *Mineral. Petrol.* **2012**, *105*, 85–111.

28. Tombros, S.; Seymour, K.St.; Williams-Jones, A.E. Controls on Tellurium in Base, Precious, and Telluride minerals in the Panormos bay Ag-Au-Te Deposits, Tinos Island, Cyclades, Greece. *Econ. Geol.* **2010**, *105*, 1097–1111.

29. Wellstead, J. Finding Tellurium. Available online: http://telluriuminvestingnews.com/222-finding-tellurium-production-canada-china-sweden-deer-horn-jet-gold/ (accessed on 11 October 2012).

30. Brown, M.; Lazo, F.; Carter, P.; Goss, B.; Kirwin, D. The geology and discovery of the Merlin Mo-Re zone of the mount Dore deposit, mount Isa Inlier, NW Queensland, Australia. *SGA News* **2010**, 9–15.

31. Voudouris, P.; Melfos, V.; Moritz, R.; Spry, P.G.; Ortelli, M.; Kartal, T. Molybdenite occurrences in Greece: Mineralogy, geochemistry and rhenium content. In Scientific Annals of the School of Geology AUTH, Proceedings of the XIX Congress of the Carpathian-Balkan Geological Association, Thessaloniki, Greece, 23–26 September 2010; pp. 369–378.

32. Melfos, V.; Voudouris, P.; Arikas, K.; Vavelidis, M. Rhenium-rich molybdenites in Thracian Mo ± Cu porphyry occurrences, NE Greece. *Bull. Geol. Soc. Greece* **2001**, *34*, 1015–1022.

33. Voudouris, P.C.; Melfos, V.; Spry, P.G.; Bindi, L.; Kartal, T.; Arikas, K.; Moritz, R.; Ortelli, M. Rhenium-rich molybdenite and rheniite (ReS_2) in the Pagoni Rachi-Kirki Mo-Cu-Te-Ag-

Au deposit, Northern Greece: Implications for the rhenium geochemistry of porphyry style Cu-Mo and Mo mineralization. *Can. Mineral.* **2009**, *47*, 1013–1036.

34. Wellmer, F.-W.; Becker-Platen, J. Sustainable development and the exploitation of mineral and energy resources: A review. *Int. J. Earth Sci.* **2002**, *91*, 723–745.

35. Driesner, T.; Pintea, I. Constraints on the conditions of wurtzite formation at the Agios Philippos Pb-Zn deposit, NE-Greece. *Berichte der Deutschen Mineralogischen Gesellschaft. Beihefte zum European Journal of Mineralogy* **1994**, *6*, 54.

36. Bonsall, T.A.; Spry, P.G.; Voudouris, P.; Tombros, S.; Seymour, K.; Melfos, V. The geochemistry of carbonate-replacement Pb-Zn-Ag mineralization in the Lavrion district, Attica, Greece: Fluid inclusion, stable isotope, and rare earth element studies. *Econ. Geol.* **2011**, *106*, 619–651.

37. Stamatakis, M.G. Phosphate deposits of Neogene age in Greece: Mineralogy, geochemistry and genetic implications.*Chem. Erde-Geochem.* **2004**, *64*, 329–337.

38. Humphries, M. *Rare Earth Elements: The Global Supply Chain*; Congressional Research Service: Washington, DC, USA, 2012.

39. Tirone, J. Greek, Nordic rare earths could save European industry. Available online: http://www.bloomberg.com/news/2012-04-26/greek-nordic-rare-earths-could-save-european-industry.html (accessed on 11 October 2012).

40. Perissoratis, C.; Moorby, S.A.; Angelopoulos, I.; Cronan, D.S.; Papavasiliou, C.; Konispoliatis, N.; Sakellariadou, F.; Mitropoulos, D. Mineral concentrations in the recent sediments of Eastern Macedonia, Northern Greece: Geological and geochemical considerations. In *Mineral Deposits within the European Community*; Boissonnas, J., Omenetto, P., Eds.; Springer: Berlin, Germany, 1988; pp. 530–552.

41. Pergamalis, F.; Karageorgiou, D.E.; Koukoulis, A.; Katsikis, I. Mineralogical and chemical composition of sand ore deposits in the seashore zone Nea Peramos—Loutra

Eleftheron Kavala (N. Greece). *Bull. Geol. Soc. Greece* **2001**, *34*, 845–850.

42. Pergamalis, F.; Karageorgiou, D.E.; Koukoulis, A. The location of Tl, REE, Th, U, Au deposits in the seafront zones of Nea Peramos-Loutra Eleftheron area, Kavala (N. Greece) using γ radiation. *Bull. Geol. Soc. Greece* **2001**, *34*, 1023–1029.

43. Vaxevanopoulos, M. Aristotle University of Thessaloniki, Thessaloniki, Greece. Personal communication, 2012.

44. Economou-Eliopoulos, M.; Eliopoulos, D. Palladium, platinum and gold concentration in porphyry copper systems of Greece and their genetic significance. *Ore Geol. Rev.* **2000**, *16*, 28–40.

45. Kiousis, G.; Economou-Eliopoulos, M.; Paspaliaris, I.; Mitsis, I. Gold, palladium and platinum recovery, as by-products, from the Skouries porphyry Cu-Au deposit, Chalkidiki area, northeastern Greece—preliminary results. In Mineral Deposit Research: Meeting the Global Challenge, Proceedings of the Eighth Biennial SGA Meeting, Beijing, China, 18–21 August 2005; pp. 991–994.

46. Economou-Eliopoulos, M. Platinum-group element distribution in chromite ores from ophiolite complexes: implications for their exploration. *Ore Geol. Rev.* **1996**, *11*, 363–381.

47. Tsoupas, G.; Economou-Eliopoulos, M. High PGE contents and extremely abundant PGE-minerals hosted in chromitites from the Veria ophiolite complex, Northern Greece. *Ore Geol. Rev.* **2008**, *33*, 3–19.

48. Kapsiotis, A.; Grammatikopoulos, T.A.; Tsikouras, B.; Hatzipanagiotou, K. Platinum-Group mineral characterization in concentrates from high-grade PGE Al-rich chromitites of Korydallos area in the Pindos ophiolite complex (NW Greece). *Resour. Geol.* **2010**, *60*, 178–191.

49. Michailidis, K.; Tarkian, M.; Bandi, A. Platinum group elements and gold (PGEs + Au) geochemistry of chromite ores from the western Chalkidiki ophiolite complex, Greece. In Proceedings of the 2nd Conference of the

Committee of Economic Geology, Mineralogy and Geochemistry of the Geological Society of Greece, Thessaloniki, Greece, 7–9 October 2005; pp. 219–228.

3
CHAPTER

Global Production Estimation of Rare Earth Elements and Their Environmental Impacts on Soils

Georgios Charalampides, Konstantinos I. Vatalis

Department of Environmental Engineering and Antipollution Control, Technological Educational Institute of Western Macedonia (TEI), Kila Kozani, Greece

ABSTRACT

Due to high demand and limited availability of rare earth elements (REEs), Europe is unable to meet its industrial needs, especially High Tech needs, today for the manufacturing sector. Therefore, the EU has included them in the group of 14 critical minerals. China currently controls completely the mining activity, the enrichment technologies and metallurgy, and end-metal products of rare earths, resulting both Europe and the U.S.A. in full industrial dependency. The exploitation and wide use of REEs in fertilizers have led to accumulation of these elements in soils, resulting in an agriculture field pollution, which affects directly the soil microfauna with a toxic potential effect.

Keywords: Rees, Production, Reserves, Environment, Fertilizers

1. INTRODUCTION

Cerium and neodymium, as major elements and praseodymium, samarium, europium, dysprosium, gadolinium, terbium, holmium, erbium, thulium, ytterbium and lutetium as subordinate amounts (including even lanthanum and often yttrium) comprise the REEs, which are also considered as critical elements due to their high tech industrial applications. Commonly, REE-bearing minerals are found scattered in a variety of forms, mostly as accessory minerals in igneous and metamorphic rocks, such as granites, gneisses and pegmatites [1] . The abundance of Ce is almost the same as environmentally much more studied elements, such as Cu and Zn; the most scarce lanthanides, Lu and Tm are actually more abundant in the Earth's crust than Cd and Se. No other group of elements in the periodic system displays such a great similarity as the REEs. They all usually form trivalent cations, though the divalent or the tetravalent oxidation state is known for most of them in chemical compounds. The effective ionic radius of the trivalent ions decreases gradually from La^{3+} (1.031 Å) to Lu^{3+} (0.861 Å) [2] .

The REEs are found, usually several together, in a variety of accessory minerals, such as phosphates, carbonates, fluorides and silicates. They rarely form more continuous ore bodies.

REEs do not occur as native elemental metals in nature, only as part of the host mineral's chemistry. For this reason, the recovery of Rare Earth Metals (REMs) must be accomplished through complex processing methods to chemically break down the minerals containing the REEs.

Despite the abundance of more than 200 known REE-bearing minerals, only three of them are considered to be the principal REE mineral ores most feasible for the extraction of REMs: bastnäsite, xenotime, and monazite [2] .

About 17 REMs are usually found together, mostly in bestnäsite with iron in Bayan Obo (Inner Mongolia), monazite and xenotime ores with radioactive thorium (Mountain View, California and Kerala, India), and iron-adsorption clays [3] [4] . More than 95% of the rare earths occur in betstnäsite, monazite and xenotime [5] . Betstnäsite contains about 70% - 75% rare earth oxides (REOs), monazite 65% - 70% and xenotime 61% - 67% [6] . Betstnäsite ores [(Ce, La, Nd)

CO_3F)] are fluorocarbonates of cerium, lanthanum, neodymium ant other REMs. Betstnäsite is the least problematic source of REOs, as it hardly contains any radioactive thorium [6] [7] .

Rare earth elements (REEs) have previously been used as geochemical tracers in groundwater flow systems owing to the commonly reported similarities between their input-normalized REE patterns and those of the aquifer rocks/sediments through which they flow and react [8] -[11] .

Rare earth elements (REEs) have been used in fertilizers in the agriculture of China for about 20 years. They have been shown to be beneficial elements for plants. For example, they have improved the yield and quality for several kinds of crops. Some studies have focused on the effects of REEs on metabolic nutrients, photosynthesis and stress resistance of plants. Other studies have shown that the environmental behaviours of REEs in soil are dominated by their low solubility. Fluorides, carbonates, phosphates and hydroxides may form neutral complexes containing REEs with a low solubility [12] .

China has the largest reserves of REEs in the world (80%) and is a major producer of REEs for the world market. Since 1990, REE fertilizer has been widely used in more than 20 Chinese provinces [13] .

The environmental behavior of REEs in soil is dominated by their low solubility (Shuzhen Zhang and Xiao-quan, 2001). Fluorides, carbonates, phosphates and hydroxides may form complexes with neutral REEs with low solubility, resulting in low dissolved concentrations in the aqueous phase of ecosystems. In solution, REEs may be complexed with inorganic ligands (e.g. carbonate, sulfate), organic ligands (e.g. humic and fulvic acids) and, at a high pH, with hydroxyl ions.

Rare earth elements (REEs) are not particularly rare in terms of abundance, but for many years remained rarely separated from each other owing to their similar chemical characteristics [14] . Although REEs (which are also known as the lanthanide series in the periodic table of elements) are widely distributed geographically, they are chiefly mined, concentrated, and separated in China.

Talens Peiro and Villalba Mendez [15] focused essentially on the chemical beneficiation of REEs for each type of mineral ore:

bestnäsite, monazite and xenotime. Their work was an attempt to estimate the material and energy requirements for the production of REM based on the theoretical chemical reactions and thermodynamics. It was indicated that material and energy balances should be of use in life cycle analysis, resource accounting, and other industrial ecology tools used to quantify the environmental consequences of meeting REMs demand for new technology products.

As concerning the REEs occurrences in Greece, significant reserves of these elements are found in alluvial deposits in the coastal and groundwater environment of Strymona gulf between Strymona River and Kavala region, Northern Greece. Specific ore deposits research estimates ore reserves of 485 million t with an average content of 1.7% rare earths. According to Eliopoulos et al. [16] , REEs are observed in Greece in three types of mineralizations: 1) in laterite/bauxite ore bodies, Central Greece; 2) in black sands derived from plutonic rocks, in the area of Strymonikos Gulf, Northern Greece and 3) in granites and granitic pegmatites of Samothrace island, Northeastern Aegean Sea. However, detailed and systematic investigations show that these bauxites and lateritic bauxites of Central Greece contain representative concentrations, ranging overall from 3275 - 6378 g/t REE. It must also be stressed that there is still an emerging interest for the red mud from the aluminum metallurgy [17] .

This paper reviews the current literature on world reserves and production of REEs and studies on using them as fertilizers and their environmental impact.

2. WORLD RESERVES AND PRODUCTION OF RARE EARTH OXIDES

The global Rare Earth Reserves by country are given in Figure 1. The reserve is defined by the USGS [18] as "the part of the reserve base which could be economically extracted or produced at the time of determination". On the contrary, the reserve base not only comprises the resources that are currently economic (= reserves) but also marginally economic reserves, and some of those that are currently sub-economic.

It is to be expected that both the reserve base and the reserve will increase in the years ahead because the steep increases in REOs prices lead to the exploration of new deposits. For example, the Chinese Ministry of Commerce announced in October 2010 that a new large rare earth deposit was found in Central China [19] . Figure 2 displays the top countries in Rare Earth mine production worldwide for the period 2010-2013 [20] .

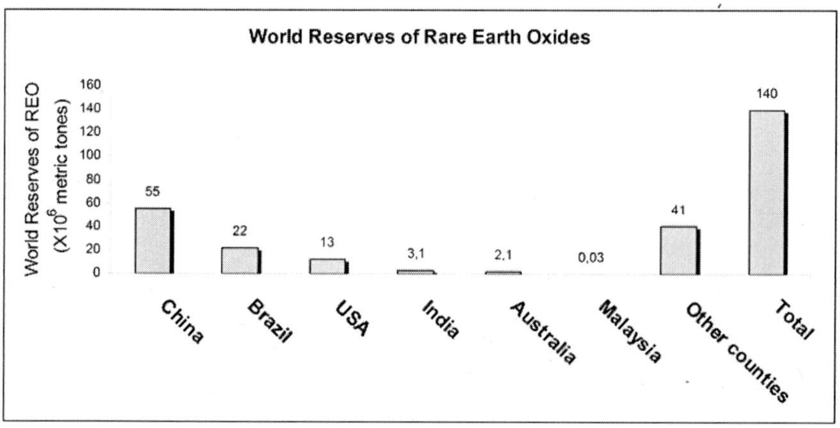

Figure 1. The possible global rare earth reserves, by country, as they are estimated by [20] .

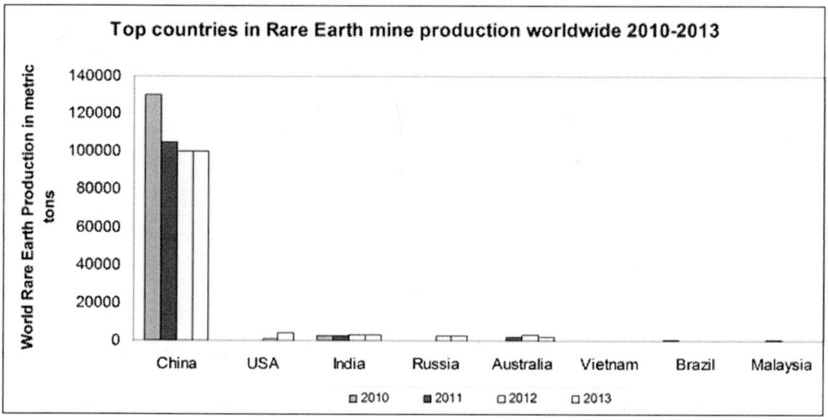

Figure 2. Top countries in rare earth mine production worldwide during 2010-2013 (after [20] & http://www.statista.com /statistics/268011/top-countries-in-rare-earth-mine-production [22]).

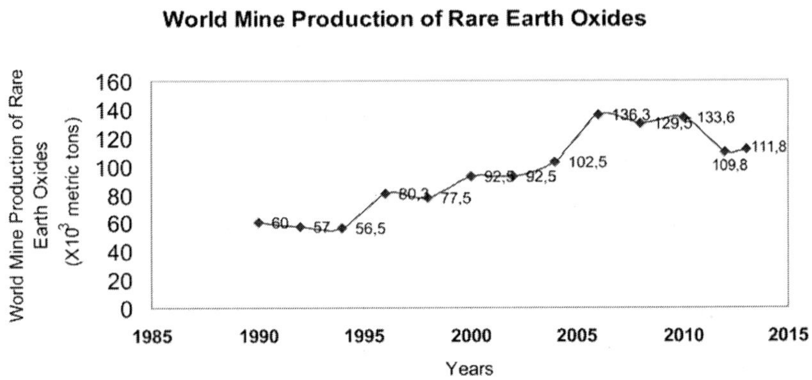

Figure 3. World mine production of REO, as it was estimated by US geological survey (modified after [18] and [20]).

On the other hand the trend of world production of RRE was increasing in recent decades, as shown in Figure 3 [18] [20] as compared the figures of the period 1990 to 2013, where the main contribution comes from China, which holds about 90% of world production, reaching 2013 to 111,800 t. The figure points out the steady increase in the rare earth production due mainly to continuous increase of the Chinese market share, particularly since 2002, when the American mine was closed due to environmental problems and low competitiveness because of low Chinese prices [21] . Though the fact that China produces more than 95% of the global production, their share of the reserves is much lower at 38%. Large deposits are also found in the USA, Australia and states of the former Soviet Union.

China continued efforts to restrict the supply of REOs and consolidate its rare earth industry. China's rare earth production and export quotas for 2013 were 93,800 tons and 31,000 tons, respectively.

In 2010, about 54,000 tons of rare earths were produced from bestnäsite in iron ores, 12,000 tons from bestnäsite, 11,000 from monazite and xenotime and 37,000 from iron-adsorption clays, which totaled almost 114,000 tones [15] . There is a first group of REMs produced in annual amounts greater than 10,000 tons (cerium, lanthanum, neodymium and yttrium), a second group that includes praseodymium and dysprosium, both of which are produced in thousands of tons. Metals produced in amounts lower than 10,000 tons can be regarded as by-products of lanthanum, cerium, neodymium and yttrium.

Generally, there are about 34 countries with rare earth deposits widely distributed in the world. In this sense they are not very rare. It is believed that in future more rare earth deposits will be discovered.

The physical and chemical beneficiation of bestnäsite is different from that of monazite and xenotime, as bestnäsite is a carbonate-fluorite mineral while monazite and xenotime are both phosphate minerals. The separation of rare earth elements poses many difficulties due to the great similarity of their chemical properties. Bestnäsite is separated from monazite due to the difference of their specific gravities using a shaking table.

3. THE ENVIRONMENTAL IMPACT OF REES AS FERTILIZERS

During the past two decades, geochemistry of REEs has been widely used in the studies of groundwater systems and showed that REE abundances and their characteristics (Ce and Eu anomalies, fractionation of light and heavy REEs etc.) in groundwater systems were mainly controlled by the rocks through which they flowed over, pH, redox conditions, solution chemistry, organic and/or inorganic complexation, colloidal and particulate matter transport [23] . However, using REEs in the future it will provide additional information about the water-rock interactions in deep seated environment and for building the foundation of water source discrimination.

Recently, the environmental pollution of rare earth elements (REE) has increased dramatically as a result of wide application of REEs in agricultural production as microelement fertilizers as well as in industry. Therefore, the biogeochemical cycling of REEs and their ecological and environmental effects are greatly concerned by many researchers. Rock weathering is an important REE source for an ecological system and a key process of global biogeochemical cycling of REE [24] . REE enrichment characteristics and mechanism analysis reveal that the released REE from weathering of carbonate rocks (mainly limestone), sillicalite and shale are mainly stored in profiles.

In addition to their use in industry, REEs are more widely applied to cropland as microelement fertilizers due to their abilities to increase yields and improve qualities of crops. As a result, more and more REEs are moving into the ecosystems. They accumulate in soils, bioaccumulate in crops, and enter the food chain, causing a problem of REE environmental pollution special to China. In recent years, REE environmental effects have become of great concern [25] . Some work should be done to and a method should be developed that can reliably estimate bioavailability of REEs to plants and thereby evaluate the potential health risk of REEs in soils and predict their impact on the ecosystem.

Recent great improvements in more routine analytical technique, the use of REEs as fertilisers, at least in East Asian agriculture, and the importance of these elements as indicators in both pedological and physiological processes and reactions have contributed to an increased interest in these previously less considered elements in environmental sciences. It is well known that the different fractions of soils metals vary considerably in their chemical reactivity and bioavailability [26] .

Several interactions between REEs and biological systems are known. Many studies in Chinese agricultural science have suggested that low concentrations of REEs may promote growth and productivity of several crops. Application of these elements, either to the seed or to the crop biomass is nowadays widely practised in Chinese agriculture, thus in a considerable part of the earth's cultivated soils. The physiological and ecophysiological mechanisms underlying their reactions have recently been given much attention.

Mixtures of REEs in fertilisers are nowadays widely used in Chinese agriculture to improve crop nutrition and have been in common use for about 20 years [27] . In most cases, however, La, several lanthanides, and probably Y are contained in the mixtures, making it difficult to conclude whether any single element is of particular importance or whether the effects are less element-specific, which seems most likely. The capacity of rare earth uptake by the roots might be limited, mainly due to the self-protection mechanisms of the plants against adverse effects of rare earths [28] .

Approximately 80% of the world's resources of rare earths are found in China which is a major producer of them for the world market [29] . Considerable work has shown that the application of rare earths with suitable dosages can improve crop growth and production, as well as resistance to poor growing conditions.

The use of REEs in various industries, especially their wide use in agriculture, has gradually been observed as a potential environmental problem in China [30] . Most of the REEs applied as seed dressing or spray on growing crops will finally reach the soil. But the amounts applied annually are certainly small compared to the total amounts present in the soil. Even if all REEs would end up and remain in the soil, thousands of years are needed to double the natural total soil pool at the recommended rates of application. It has to be considered, however, that REEs supplied as fertilisers are much more soluble and reactive than the average soil pool and therefore of more immediate environmental interest.

The results of several studies [28] [30] showed that the accumulation of REEs in soil was significant in the study area and its concentration was strongly correlated with the distance from the pollution source.

Gillet and Ponge [31] have indicated that soil fauna is very sensitive to the changes in the soil environment. The soil fauna community responds to soil degradation and soil heavy metals pollution, such as Cd, Cu, Pb and Hg, in many ways. Thus, soil fauna can be an indicator of soil degradation and soil pollution, and is often used to evaluate soil quality.

Heavy metal pollution of soil can lead the soil fauna community to change, but the response of different soil fauna to contamination is different. Almost all the deposits of REE contain the radioactive element thorium, and therefore environmental cost should be related to market competitive force.

4. DISCUSSION

Due to high demand and limited availability of REEs, Europe is unable to meet today its industrial needs for the manufacturing sector. Therefore the EU has included them in the group of 14 critical minerals. Namely, China currently controls completely the mining activity (almost 95% of world production) and thereafter the production, the enrichment technologies and metallurgy and end-metal products of rare earths, resulting for both Europe and the U.S.A. in full industrial dependency, reaching almost to 100%. This is happening at a time when the demand and needs tend globally constantly growing at an annual rate of 8% - 11%. There are large differences in market prices of rare earth oxides. Also, the prices depend on the degree of purity determined by the specifications in the applications. The price ranges from 11 $/kg La_2O_3 to 1600 $/kg Eu_2O_3(www.lynascorp.com [32]). However, it is of importance to emphasize that illegal trading practices are often used in China to circumvent direct control of important secondary raw material flows, including recycled REEs. Preliminary statistics of June 2014 show illegal smuggling hidden among 67 REEs recycling projects across the country. MIIT urged the government to regulate REE recycling [33] .

While recovering the world economy from financial crisis and the boom of rare earth market, it seems to be very interesting and profitable to invest to rare earth industry. According to Chegwidden and Kingsnorth [34] and Watanabe [35] , there are about 200 projects on rare earth exploration, under preparation and exploitation. Among so many projects, about 25 of them are promising to be the competitors of future multi-supply.

China government's decision to subsidize REE producers pursuing technological upgrades, shows the commitment of the country to address energy efficiency and environmental issue and to promote value added products [36] .

The most promising producers out of China which could get into production recently should be MolyCorp, Lynas, Simelt of Estonia, Great Western Group, and three Japanese joint ventures in Vietnam, India and Kazakhstan. The total production capacity out of China after 2015 should be more than 170,000 tons [37] .

The EC report from on the Implementation of the Raw Materials Initiative' of June 24, 2013 [38] confirmed that by 2020 actions would include: setting up innovative pilot projects, including demonstration plants on exploration, mining and processing; finding substitutes for raw materials in short supply; creating a network of research centers on sustainable raw materials management; using EU standardized instruments for surveying resources; creating a 3D geological map; modeling raw material trends; undertaking complete lifecycle assessment of EU raw materials needs.

5. CONCLUSIONS

Almost 50,000 tons of rare earths are needed per year out of China at present time. With a rate of 15% demand increase, then in 2015, the total demand out of China should be at least 80,000 tons on the basis of the world economy continuous increase. The products of exploitation could be rare earth concentrate and compounds like oxides, metals and alloys, depending on target consumer and the rare earth contents in the deposits.

The distribution of metals in the aerial parts of plants is the result of both the fixation and the transfer mechanisms. The complexation of organic ligands in the xylem vessels plays an important role, and elements with strong complexation capacities (strong motilities) are preferentially transported to the top. The REEs concentrations, as well as total nitrogen, total potassium, and pH, had some effect on the soil macrofauna community. A small amount of REEs can promote soil macrofauna diversity, but a large amount of REEs can reduce the soil macrofauna diversity.

To fully understand the effects of REEs in agricultural application, environment and human health, it is suggested that future research in several areas is needed, which will shed new light on the effects of REEs on agriculture, environment and human health. First, the dynamic changes and fates of REEs used in fertilizers should be thoroughly investigated. Second, the indirect effects of REEs on human health and ecosystems through their use in agriculture should be explored. Third, it is necessary to make an assessment of REEs as hazards in their use in agriculture and to develop standards for soil quality.

REFERENCES

1. Greenwood, N.N. and Earnshaw, A. (1984) Chemistry of the Elements. Pergamon Press, Oxford.
2. Tyler, G. (2004) Ionic Charge, Radius, and Potential Control Root/Soil Concentration Ratios of Fifty Cationic Elements in the Organic Horizon of a Beech (Fagus sylvatica) Forest Podzol. Science of the Total Environment, 329, 231-239.
3. Naumov, A.V. (2008) Review of the World Market of Rare-Earth Metals. Russian Journal of Non-Ferrous Metals, 49, 24.
4. U.S. Geological Survey (USGS) (2011) Rare Earth Elements-End Use and Recyclability, Scientific Investigations Report 2011-5094, U.S. Department of the Interior.
5. Gupta, C.K. and Krishnamurthy, N. (1992) Metallurgy of Rare Earths. International Materials Review, 37, 197.
6. Gupta, C.K. and Krishnamurthy, N. (2005) Extractive Metallurgy of Rare Earths. CRC Press, Boka Radon.
7. Hambel, C.A., Ed. (1971) Rare Metals Handbook. 2nd Edition, Robert E. Krieger Publishing Co. Inc., Malabar, 715.
8. Banner, J.L., Wasserburg, G.J., Dobson, P.F., Carpenter, A.B. and Moore, C.H. (1989) Isotopic and Trace Element Constraints on the Origin and Evolution of Saline Groundwaters from Central Missouri. Geochimica et Cosmochimica Acta, 53, 383-389.
9. Smedley, P.L. (1991) The Geochemistry of Rare Earth Elements in Groundwater from the Carnmenellis Area, Southwest England. Geochimica et Cosmochimica Acta, 55, 2767-2779.
10. Johannesson, K.H., Cortés, S.A., Ramos, L.J.A., Ramírez, G.A. and Durazo, J. (2000) "Rock-Like" versus "Seawater-Like" REE Signatures of Groundwaters. Geological Society of America. Abstracts with Programs, 32, A-188.

11. J Janssen, R.P.T. and Verweij, W. (2003) Geochemistry of Some Rare Earth Elements in Groundwater, Vierlingsbeek, The Netherlands. Water Research, 37, 1320-1350.

12. Zhang, S.Z. and Shan, X.Q. (2001) Speciation of Rare Earth Elements in Soil and Accumulation by Wheat with Rare Earth Fertilizer Application. Environmental Pollution, 112, 395-405.

13. Yu, Z. and Chen, M., Eds. (1995) Rare Earth Elements and Their Applications. Metallurgical Industry Press, Beijing, 286-294.

14. Cindy, H. (2010) China's Rare Earth Elements Industry—What Can the West Learn? Institute for Analysis of Global Security, Washington DC, 42 p.

15. Laura, T.P. and Gara, V.M. (2013) Material and Energy Requirement for Rare Earth Production. JOM, 65, 1327-1340.

16. Demetrios, E., George, E., Ioannis, T. and Christos, P. (2014) The Potential of Rare Earth Elements in Greece. Proceedings of the 1st European Rare Earth Resources Conference, Milos, 4-7 September 2014, 308-316.

17. Arvanitidis, N. and Papavasileiou, K. (2011) Mineral Wealth. Institute of Geology and Mineral Exploration of Greece, Athens.

18. USGS (2010) Mineral Commodity Summaries. US Geological Survey, Reston, VA.

19. MOFCOM (2010) The Ministry of Commerce of the People's Republic of China: Large Rare Earths Deposit Found in Central China.

20. USGS (2014) Mineral Commodity Summaries. US Geological Survey, Reston, VA, 129-130. [Paper reference:5]

21. The Greens/EFA Group in the European Parliament, Final Report for Darmstadt (2011) Study on Rare Earths and Their Recycling.

22. http://www.statista.com/statistics/268011/top-countries-in-rare-earth-mine-production/.

23. Sun, L.H., Gui, H.R. and Chen, S. (2011) Rare Earth Element Geochemistry of Groundwaters from Coal Bearing Aquifer in Renlou Coal Mine, Northern Anhui Province, China. Journal of Rare Earths, 29, 185.

24. Song, Z.L., Liu, C.Q., Han, G.L. and Wang, Z.L. (2006) Enrichment and Release of Rare Earth Elements during Weathering of Sedimentary Rocks in Wujiang Catchments, Southwest China. Journal of Rare Earths, 24, 491-496.

25. Cao, X.D., Wang, X.R. and Zhao, G.W. (2000) Assessment of the Bioavailability of Rare Earth Elements in Soils by Chemical Fractionation and Multiple Regression Analysis. Chemosphere, 40, 23-28.

26. Davies, B.E. (1992) Inter-Relationships between Soil Properties and the Uptake of Cd, Cu, Pb and Zn from Contaminated Soils by Radish (Raphanus sativus L.). Water, Air, & Soil Pollution, 63, 331-342.

27. Pang, X., Li, D.C. and Peng, A. (2002) Application of Rare-Earth Elements in the Agriculture of China and Its Environmental Behavior in Soil. Environmental Science and Pollution Research, 9, 143-148.

28. Diatloff, E., Smith, F.W. and Asher, C.J. (1976) Rare Earth Elements and Plant Growth: I. Effects of Lanthanum and Cerium on Root Elongation of Corn and Mungbean. Journal of Plant Nutrition, 18, 1963-1976.

29. Brown, P.H., Rathjen, A.H., Graham, R.D. and Tribe, D.E. (1990) Rare Earth Elements in Biological Systems. In: Gschneidner Jr., K.A. and Eyring, L., Eds., Handbook on the Physics and Chemistry of Rare Earths, Volume 13, Elsevier, Amsterdam, 423-452.

30. Hu, X., Ding, Z.H., Chen, Y.J., Wang, X.R. and Dai, L.M. (2002) Bioaccumulation of Lanthanum and Cerium and Their Effects on the Growth of Wheat (Triticum aestivum L.) Seedlings. Chemosphere, 48, 621-629.

31. Gillet, S. and Ponge, J.-F. (2006) Are Acid-Tolerant Collembola Able to Colonise Metal-Polluted Soil. Applied Soil Ecology, 26, 219-231.

32. www.lynascorp.com.

33. Torrisi, A. (2014) China's Ministry of Industry and Information Technology Calls for Regulations on Rare Earths Recycling. Industrial Minerals Magazine, 13 June 2014.

34. Chegwidden, J. and Kingsnorth, D.J. (2010) Rare Earths: Facing the Uncertainties of Supply. Proceedings of the Sixth International Rare Earths Conference, Hong Kong, 9-11 November 2010.

35. Watanabe, Y. (2010) Japan's Search for Alternative Rare Earths Supply. Proceedings of the Sixth International Rare Earths Conference, Hong Kong, 9-11 November 2010.

36. Torrisi, A. (2014) China Government to Subsidise Environmental and Efficiency Upgrades of Rare Earths Companies. Industrial Minerals Magazine, July 10, 2014.

37. Chen, Z.H. (2011) Global Rare Earth Resources and Scenarios of Future Rare Earth Industry. Journal of Rare Earths, 29, 1-6.

38. Nicoletopoulos, V. (2014) European Policies on Critical Raw Materials, including REE. Proceedings of the 1st European Rare Earth Resources Conference, Milos, 4-7 September 2014.

4

CHAPTER

Analysis of Different Rare Metals, Rare Earth Elements, and other Common Metals in Groundwater of South West Bank/Palestine by ICP/MS-Data and Health Aspects

Fuad Al-Rimawi[1], Khalid Kanan[1], Mutaz Qutob[2,3]

[1]Faculty of Science and Technology, Al-Quds University, East Jerusalem, Palestine; [2]Department of Earth and Environmental Studies, Faculty of Science and Technology, Al-Quds University, East Jerusalem, Palestine; [3]Aquatic and Aquaculture Research Laboratory, Al-Quds University, East Jerusalem, Palestine.

ABSTRACT

Ground water samples are collected from south West Bank/Palestine and analyzed for different rare elements (Rb, Zr, U, P, Ti, V), rare earth elements (La, Ce, and Nd), and other common trace metals (Li, Na, Mg, Ca, Sr, Ba, K, Bi) that most of them usually have no maximum acceptable limits as either they are considered not to be toxic to human health or there is no sufficient data about their toxicity to human health. This study was conducted to determine the water quality of ground water which is used for drinking in the study area. Water samples from ten groundwater wells were obtained in three different dates of the year (November 2012, March 2013, and April 2013). Three water samples were obtained from each well for each sampling date; so a total of 90 water samples were collected from the ten wells. The results obtained from this study suggest a possible risk to the population of the study area given the high concentration of some metals that have no maximum allowed concentration, and the fact that for many people in the study area, ground water is a main source of their water supply.

Keywords: Rare Metals; Rare Earth Elements; WHO Limits; ICP/MS; Ground Water; West Bank

1. INTRODUCTION

The great importance of water for man makes the accurate management of this natural resource a priority in order to preserve its sustainability. Water analysis plays a decisive role in regulations that determine the water quality levels for specific uses e.g., for drinking water [1]. Water analysis applicability is also extended to the study of water origin and evolution, taking into account the natural and anthropic influences, e.g. water-water and water-rock interactions, groundwater exploitation regulation (protection and recharge of aquifers), contamination processes, etc. It is often assumed that natural ground water from deep wells is clean and healthy [2]. This is usually true with regards to bacteriological composition. However, a large variety of solutes are transferred to the water via complex pathways and processes, such as atmospheric and terrestrial sources. The result is the occurrence of different components in the water, which define the hydrochemical fingerprint. This fingerprint permits identification of water, the media and the interactions between them, including those associated with anthropic activities. The concept is comparable to individual fingerprint identification. High concentrations of many chemical elements can occur in ground water due to variations in the regional geology and water/rock interactions [3-6]. Inorganic chemical quality of ground water is, however, rarely adequately tested before the wells are put into production [7].

Occurrence of trace elements in ground water is very sensitive to geological changes and, in many cases, to anthropic influences [7]. Hence, the advances in analytical techniques facilitate the application of the hydrochemical fingerprint concept. Inductively coupled plasmamass spectrometry (ICP-MS) is particularly suitable for this purpose due to the possibility of rapid multi-element analysis in combination with excellent detection limits [8-11]. The ICP-MS allows one to obtain complete information on the trace elements occurring in water, including those present in sub-ppb concentrations. The analytical capability of ICP-MS in this field has led to consideration of this technique for routine water analysis.

Several studies have shown that wells in areas with particular geological features yield water that does not meet established drinking water norms without any influence from anthropogenic contamination e.g. Misund et al., 1999 [7]; Frengstad et al., 2000 [12], Reimann et al. 2003 [13]. This work is a continuation of a previous work where heavy metal contamination of the groundwater of south West Bank in Palestine was assessed by analysis of different heavy metals (Tl, Pb, Bi, Cr, Mn, Co, Ni, Cu, Zn, Mo, Al, Ag, and Cd) and results showed that 93% of all samples analyzed contained one or more of the 13 metals studied each in varying concentration, and some metals (Pb Cd, and Al) are found to be higher than the allowed WHO limits in some samples. The present paper reports analytical results for some rare elements (Rb, Zr, U, P, Ti, V, La, Ce, and Nd) as well as other common elements (Ca, Mg, K, Sr, Ba, and Bi) which most of them have no maximum allowed concentrations by WHO or any other regulation agencies. The current study also document that quite a number of elements for which no drinking water guideline values or maximum acceptable concentration limits have been established can occur at unpleasantly high levels in natural well waters (e.g. Li, Na, Mg, Ca, Sr, Ba, K, Bi).

2. EXPERIMENTAL

2.1. Study Area

West Bank is divided into eleven districts: Bethlehem, Hebron, Jenin, Jericho, Jerusalem, Nablus, Qalqilya, Ramallah and Al-Bireh, Salfit, Tubas, and Tulkarm. The districts are sub-divided into 89 municipalities, and there are local councils that manage all infrastructure and basic services in the towns and villages.

2.1.1. Geography and Geology

South of West Bank including Jerusalem, Beitlehem, and Hebron have a highly varying topography and altitude; with the highest point at 1011 m above sea level and the lowest at 150 m above sea level. In terms of geology, the majority of the study area is rocky mainly comprised of carbonate sediment such as limestone, chalky limestone and dolomite with marl. The geological rock age formations range from Turonian to Upper Cenomanian [14].

2.1.2. Climate

Water samples were obtained from south of West Bank/ Palestine including three areas: Jerusalem, Beitlehem, and Hebron. The climate of the study area ranges from arid to semi-arid with an increase in aridity towards the Negev desert in the south and the Jordan valley in the east. The monthly average temperature of the study area ranges from 7.5°C - 10°C in the winter to 22°C in the summer. The minimum temperature is −3°C in January and the maximum is 40°C in August. Most of the rain falls during December through February, although there may be rain from mid-October to the end of April. Water shortage is a serious problem facing the study area due to the arid and semi-arid climatic conditions [14].

2.1.3. Groundwater: Aquifers and Wells

The West Bank lies over the Mountain aquifer which is divided into the eastern aquifer, the northeastern aquifer, and the western aquifer. The eastern aquifer and part of the northeastern aquifer flow east towards the Jordan River. The western aquifer and part of the northeastern aquifer flow westerly towards the Mediterranean Sea [15, 16]. In the study area, there are 24 wells [17].

2.2. Sampling and Analysis

Ten groundwater wells were selected as representative wells of the study area. **Figure 1** shows location of the wells in the study area analyzed in this study. Three samples were obtained from each well. The selected wells were sampled three times in three different dates of 2012 and 2013 (November 2012, March 2013, and April 2013). A total of 90 water samples were collected from the ten wells. The water samples were collected in 1-liter high density polyethylene bottles (pre-cleaned with 10% nitric acid followed by repeated rinsing with bi-distilled water), stabilized with ultrapure nitric acid (0.5% HNO_3), preserved in a cool place (about 4°C) and transported to the lab of Al-Quds University for further analysis. The samples were then analyzed for trace elements content by ICP/MS (Agilent technologies 7500 series).

For accurate quantitative determination of trace metals in water samples, an internal standard method was used using Y as internal standard and a multi-standard calibration method: (22 metals standard (Ag 10 ppm, Al 50 ppm, B 50 ppm, Ba 10 ppm, Bi 100 ppm, Ca 10 ppm, Cd 10 ppm, Co 10 ppm, Cr 50 ppm, Cu 10 ppm, Fe 10 ppm, K 100 ppm, Li 50 ppm, Mg 10 ppm, Mn 10 ppm, Mo 50 ppm, Na 50 ppm, Ni 50 ppm, Pb 100 ppm, Sr 10 ppm, Tl 50 ppm, Zn 10 ppm, matrix 5% HNO_3). Other standards containing Rb, Zr, U, P, Ti, V, La, Ce, Nd with concentration of 1000 ppm in 5% HNO_3 solution. Samples were prepared by dilution of 1.0 mL of the water samples to 10.0 mL with 0.3% ultrapure nitric acid and analyzed by ICP/MS. Each sample was analyzed three times and the results are expressed as mean ± SD (SD: standard deviation). Relative standard deviation (RSD) of the three results are calculated and found to be less than 5% for all samples for all metals analyzed in this study, reflecting

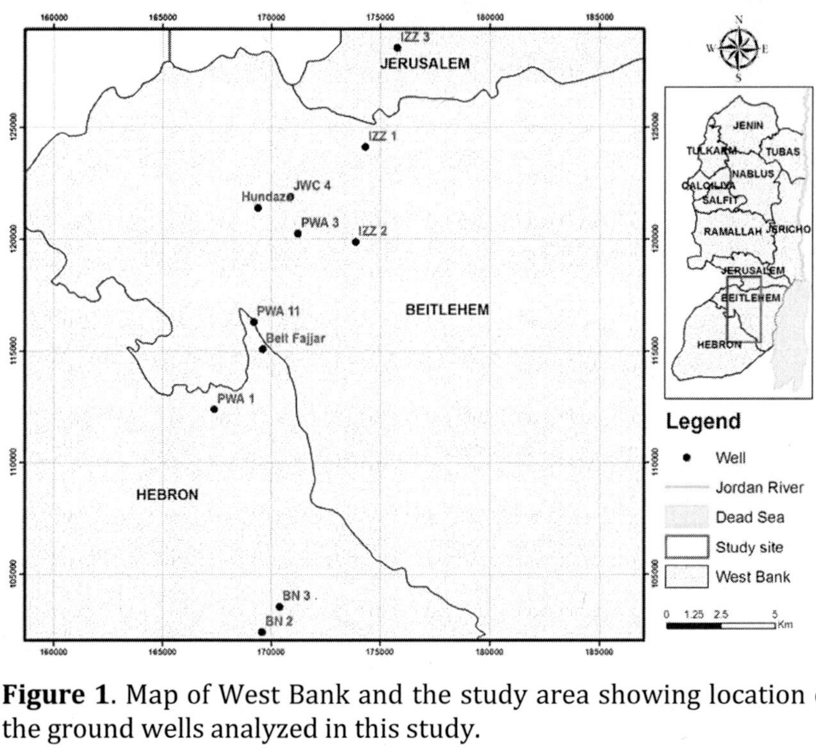

Figure 1. Map of West Bank and the study area showing location of the ground wells analyzed in this study.

the precision of the method for the analysis of these heavy metals. Calibration curves for all metals analyzed were constructed by plotting the ratio of the intensity of the analyte metal to that of the internal standard (Y) vs. concentration of the trace metal (in ppb), and results showed that the calibration curves are linear with correlation coefficient (r^2) greater than 0.999 for the trace metals analyzed.

The Agilent Technologies 7500 Series ICP-MS (Agilent 7500) can measure trace elements as low as one part per trillion (ppt) and quickly scan more than 70 elements to determine the composition of an unknown sample with a MassHunter Workstation software automates the analysis and accurately interprets the resulting data. The ICP/MS instrument consists of an on-board peristaltic pump that controls the flow of sample solution into and waste (drain) out of the instrument, a nebulizer (Micro Mist nebulizer) that uses a stream of argon to disperse the sample, an ICP Argon plasma torch using Argon as plasma gas, auxillary gas and nebulizer (carrier) gas, two pumps for evacuationquadrupole mass analyzer with 0.8 amu resolution at 10% height, an octapole reaction system (ORS), and electron multiplier detector.

The operating conditions are as follows: nebulizer gas (argon) flow rate: 0.9 L/min, auxiliary gas (argon) flow 0.3 L/min, plasma (Argon) gas flow: 15 L/min, reaction gas flow (helium) 4 mL/min, lens voltage 7.25 V, ICP RF power: 1100 W, CeO/Ce = 1%, and Ce^{+2}/Ce^{+1} = 1%.

2.3. Statistical Analysis

Data were analyzed using Origin 9 software. Statistical differences were tested using one way ANOVA. Differences were considered significant at p values ≤ 0.05.

3. RESULTS AND DISCUSSION

3.1. Concentration of the Analyzed Elements

3.1.1. Rare Elements

The detection limits of the studied elements are reported in **Table 1**, and are three times the standard deviation estimated from six replicate measurements of the blank (1% v/v HNO_3). The detection limits by ICP-MS range between 0.01 and 8.0 ppb for the trace elements analyzed in this study. **Table 1** summarizes concentrations (minimum, median, maximum, average, and standard deviation) of the trace elements (Li, Na, Mg, Ca, Sr, Ba, Bi, K which are analyzed on November 2012, March 2013 and April 2013) as well as rare elements (Rb, Zr, La, Ce, Nd, U, P, Ti, V which are analyzed in April 2013) for the groundwater samples analyzed in this study. It provides additional information on water standards (WHO, European Union, US EPA) for some elements analyzed in this study (Na, Ba, U, and Sr). One of the largest differences in opinion between the authorities exists for U, for which the EU has not defined any maximum acceptable limit, while WHO (1998) has set a limit of 2 ppb, based on toxicological considerations [18]. The concentration of U in the water samples for the 10 ground wells analyzed in this study was found to be in the range of 0.81 to 1.48 ppb with a median of 0.94 and average of 0.98 ppb. Although all water samples have U within the acceptable limits, it has been detected in all samples analyzed in this study. Uranium is present in the environment as a result of leaching from natural deposits, release in mill tailings, emissions from the nuclear industry, the combustion of coal and other fuels and the use of phosphate fertilizers that contain uranium. Its presence in drinking water is most commonly from natural sources. Uranium occurrence in groundwater has been already reported in many regions; in a survey of 130 sites (approximately 3700 samples) in Ontario, Canada, conducted between 1990 and 1995, the mean of the average uranium concentrations (range 0.05 - 4.21 μg/litre) in treated drinking-water was 0.40 ppb [19]. A study in Finland examined a population receiving drinking-water containing uranium with a median concentration of 28 ppb [20]. In a study of 476 Norwegian groundwater samples, 18% had uranium

Table 1. Concentration (in ppb) of different elements detected in groundwater wells analyzed in this study.

Element	LOD (ppb)	Sampling date	Minimum (ppb)	Median (ppb)	Maximum (ppb)	Average (ppb)	SD	Limits
		November 2012	0.7	2.10	3.50	2.06	0.84	
Li	0.1	March 2013	0.99	2.33	12.4	3.17	2.91	No limits
		April 2013	1.09	1.86	5.31	2.32	1.29	
		November 2012	115.4	138.0	183.1	139.1	19.6	
Na	3	March 2013	157.1	339.5	577.7	356.5	104.0	200 ppm (EU Directive). *20 ppm (desirable limit according to WHO 2004).
		April 2013	24405	21112	57302	21011	13224	
		November 2012	396.0	487.4	581.3	480.9	59.2	
Mg	5	March 2013	198.4	6376.7	7459.0	5864.5	1800.0	No limits
		April 2013	21708	26123	32211	26001	2501	
		November 2012	71.0	99.3	121.6	96.9	15.4	
Ca	5	March 2013	164.7	590.2	717.7	555.7	134.4	No limits
		April 2013	9901	11405	13308	11621	1011	
		November 2012	74.6	232.6	831.6	349.8	241.7	
Sr	1	March 2013	147.3	283.4	623.6	327.8	173.7	4 ppm (U.S. EPA)
		April 2013	189.8	272.0	720.9	376.7	204.4	
		November 2012	43.6	81.8	165.9	99.4	44.1	

Ba	1.0	March 2013	7.2	16.53	58.5	19.14	13.97	700 ppb (WHO)
		April 2013	21.6	49.8	65.9	45.8	13.8	
		November 2012	3.41	18.20	215.7	48.0	69.4	
Bi	0.01	March 2013	1.62	1.63	2.8	1.71	0.32	No limits
		April 2013	0.02	0.16	1.10	0.33	0.45	
		November 2012	194.0	248.3	311.6	247.7	40.0	
K	8.0	March 2013	903.6	1227.1	1789.1	1216.6	229.0	No limits
		April 2013	1305	1612	2914	1721	460	
Rb	0.01	April 2013	0.28	0.59	1.0	0.61	0.20	No limits
Zr	0.01	April 2013	0.02	0.11	1.13	0.30	0.36	No limits
La	0.01	April 2013	0.01	0.01	0.02	0.01	0.007	No limits
Ce	0.01	April 2013	0.02	0.05	0.09	0.05	0.03	No limits
Nd	0.01	April 2013	0.01	0.01	0.03	0.01	0.01	No limits
U	0.001	April 2013	0.81	0.94	1.48	0.98	0.20	2 ppb (WHO)
P	0.1	April 2013	0.31	6.15	9.66	5.70	2.9	No limits
Ti	0.02	April 2013	0.09	0.19	0.32	0.21	0.08	No limits
V	0.03	April 2013	1.90	3.28	4.99	3.34	1.04	No limits

concentrations in excess of 20 ppb [12]. Concentrations in excess of 20 ppb have been reported in groundwater from parts of New Mexico, USA [21], and central Australia [22]. Intake through drinking-water is normally low; however, in circumstances in which uranium in present in a drinking-water source, the majority of intake can be through drinking-water [23].

No drinking water standards are defined by EU, WHO or US EPA for Rb, Zr, La, Ce, Nd, P, Ti, and V. Several of these elements have documented health effects [13]. For others, for example the rare earth elements, knowledge about health effects are rather limited [24]. However all of these elements are detected in all groundwater samples analyzed in this study. Additionally, some of these elements are unexpectedly and surprisingly detected with high concentrations e.g. P, V; the concentration of P ranges from 0.31 to 9.66 ppb with an average of 5.70 ppb, and the concentration of V ranges from 1.90 to 4.99 ppb with an average of 3.34 ppb. The concentration of Rb, Zr, La, Ce, Nd, and Ti was found to be in the range of 0.28 - 1.0 (average = 0.61 ppb), 0.02 - 1.13 (average = 0.30 ppb), 0.01 - 0.02 (average = 0.01 ppb), 0.02 - 0.09 (average = 0.05 ppb), 0.01 - 0.03 (average = 0.01 ppb), 0.09 - 0.32 (average = 0.21 ppb), respectively. In addition to the potential pollution of groundwater with these rare elements, the concentrations of these elements such Bi, and rare earth elements, which are extremely low in natural waters, are of particular interest to fingerprint the groundwater sources [25].

3.1.2. Common Elements

Concerning Na, Mg, Ca, Sr, and Li, they are elements of primarily geogenic origin. Their presence in groundwater is due to rock-water interaction over time and/or weathering of silicate and carbonate minerals in particular [13]. Of these elements, only sodium and strontium has maximum acceptable limits, while the rest (Mg, Ca, and Li) have no limits. However, several of the elements that have no maximum allowed limits in drinking water are deemed to have health effects [13]. Furthermore, there are a number of elements that are rather toxic and/or carcinogenic for which no maximum allowed limit has been defined, probably often under the assumption that "natural concentrations in water must be very low" (WHO, 1993; statement with regard to Be) [26]. Our results showed that the concentrations of some elements analyzed cover between one and three orders of magnitude. For many of these elements, little is known about health effects at drinking water concentrations. In this respect, it is important to analyze and monitor regularly such elements that have no maximum acceptable limits in drinking water.

Sodium exists in ground water naturally as a result of rock-water interactions. Sodium ion is ubiquitous in water, owing to the high solubility of its salts and the abundance of Na-containing mineral deposits [27]. According to WHO 2004 [28], the desirable limit of sodium is 20 ppm, while the limit is 200 ppm according to EU Directive regulations [29]. Our results showed that the concentration of sodium ranges from 0.12 - 0.18 (average is 0.14 ppm), 0.16 - 0.58 (average is 0.36 ppm), 24.4 - 57.3 (average is 21.0 ppm) for the water samples analyzed in November 2012, March 2013, and April 2013, respectively. These results show that all water samples collected in April 2013 exceeded the desirable limit of Na according to WHO 2004 (20 ppm) while this concentration is still within the limit of EU directive (200 ppm). Anyway, the concentration of sodium in ground water collected in April 2013 is relatively large which indicates that groundwater of the study area needs regular monitoring. Although sodium is essential to human body, excessive intake or very high doses of Na may cause acute effects such as nausea, vomiting, inflammatory reaction in the gastrointestinal tract, thirst, muscular twitching, and convulsions. For long-term lower level exposures, the health effect of primary concern is hypertension, central nervous system disturbances such as convulsions, confusion and pulmonary edema are possible [27]. Magnesium is present also naturally in ground water. There is no limit for Mg according to WHO or EU. However, regulative agencies in some countries set a limit for Mg. For example, the Bureau of Indian Standards (BIS) has set a limit of 30 ppm for Mg in drinking water [30]. As in the case of Na, the concentration of Mg in the ground water samples collected in April 2013 is high (21.7 - 32.2 ppm) and exceeded the limit of BIS for Mg in drinking water (30 ppm) for some groundwater samples (more than 60% of the analyzed water samples). Increased intake of Mg salts may cause a change in bowel habits (diarrhea), but seldom causes hypermagnesemia in persons with normal kidney function. Excess Mg concentration may lead to changes in mental status, nausea, diarrhea, loss of appetite, muscle weakness, difficulty in breathing, extremely low blood pressure, and irregular heartbeat [31].

The concentration of calcium, which present naturally in ground water, was found to be in the range of 9.9 - 13.3 ppm with an average of 11.6 ppm in April 2013, while its concentration in November 2012 and March 2013 is low (less than 1 ppm). The desirable limit of Ca is specified as 75 ppm for drinking water according to WHO [32]. Concern for excess calcium intake is directed primarily to those who are prone to milk alkali syndrome (the simultaneous presence of hypercalcemia, metabolic alkalosis and renal insufficiency) and hypercalcemia. Although calcium can interact with iron, zinc, magnesium and phosphorous within the intestine, thereby reducing the absorption of these minerals, available data do not suggest that these minerals are depleted when humans consume diets containing calcium above the recommended levels [32]. Water hardness has been related with cardiovascular diseases, though it is not clear whether it is due to calcium and/or magnesium or because metals are more soluble in hard waters [33]. Strontium which behaves chemically like calcium has a geogenic origin, so it present naturally in ground water. The concentration of Strontium in the 90 ground water samples analyzed in this study was found to be in the range of 74.6 - 831.6 (average is 349.8 ppb), 147.3 - 623.6 (average is 327.8 ppb), and 189.8 - 720.9 (average 376.7 ppb) for water samples collected in November 2012, March 2013, and April 2013, respectively. The US Environmental Protection Agency (EPA) has developed a limit of 4 ppm for strontium levels in drinking water [34]. However WHO does not set a maximum acceptable limit of Sr in drinking water. According to EPA, Water that contains more than 4 ppm of Sr should not be used for drinking water or to prepare beverages, food with water as a major ingredient or infant formulas. Lithium which exist in the groundwater naturally was found to be in the ground water samples analyzed in this study in the range of 0.70 - 3.5, 0.99 - 12.4, and 1.09 - 5.31 ppb for the samples analyzed in November 2012, March 2013, and April 2013, respectively. There is no maximum acceptable limit of Li in drinking water.

Regarding the concentration of potassium in the ground water samples analyzed in this study, it ranges from 194.0 to 311.6 ppb (with an average of 247.7 ppb), 903.6 to 1789.1 ppb (average of 1216.6 ppb), and 1305.3 to 2914.6 ppb (average 1721.3 ppb) for water samples collected in November 2012, March 2013, and April 2013, respectively. There is no limit for K in drinking water according to WHO or EPA.

However, increased exposure to potassium could result in significant health effects in people with kidney disease or other conditions, such as heart disease, coronary artery disease, hypertension, diabetes, adrenal insufficiency and pre-existing hyperkalemia. Symptoms of hyperkalemia include weak pulse rate, irregular heartbeat and nausea [35]. Infants have also a limited renal reserve and imamture kidney function and may therefore be morevulnerable [36]. The concentration of Barium in the ground water samples analyzed in this study was found to be in the range of 43.6 to 165.9 ppb (with an average of 99.4 ppb), 7.2 to 58.5 ppb (average of 19.14 ppb), and 21.6 to 65.9 ppb (average 45.8 ppb) for water samples collected in November 2012, March 2013, and April 2013, respect tively. WHO has set a maximum allowed limit for Ba in drinking water of 700 ppb [18]. Ba in drinking-water is likely to be associated with groundwater of low pH from granite-like igneous rocks, alkaline igneous and volcanic rocks and manganese-rich sedimentary rocks. Concentrations are, therefore, expected to be relatively stable [37]. When people are exposed to Ba for short periods at levels above the maximum contaminant level, they may experience gastrointestinal disturbances and muscular weakness. Additionally, Ba has the potential to cause high blood pressure when exposed to levels above the limit for long periods of time [37]. Concerning the concentration of Bismuth in the ground water samples analyzed in this study, our results showed that the it ranges from 3.41 to 215.71 ppb (with an average of 48.02 ppb), 1.62 to 2.82 ppb (average of 1.71 ppb), and 0.02 to 1.10 ppb (average 0.33 ppb) for water samples collected in November 2012, March 2013, and April 2013, respectively. Although there are no limits for Bi in drinking water by WHO, EPA or any other agency, it has been detected in all water samples analyzed in this study.

3.2. Effect of Sampling Date on the Content of the Analyzed Elements

Statistical analyses were conducted to test if there is significant difference in the concentration of the studied elements in the groundwater wells as a function of the sampling date (November 2012, March 2013, and April 2013), and results confirmed that there is a significant difference, denoting that element concentration in the wells vary significantly with sampling time. As it is seen in **Table 1**, the highest concentration of Li varies from 3.5 ppb in November 2012 to 12.4 ppb in March 2013. For Na, Mg, Ca, the concentration increases significantly from November 2012 through March 2013 and April 2013. For example, the highest concentration of Na increases from 0.18 in November 2012 to 57.3 ppm in April 2013. The same applies for Mg and Ca where the highest concentration increases from 0.58 to 32.2 (for Mg), and from 0.12 to 13.3 ppm (for Ca). For K, the same applies where the highest and lowest concentration increases from November 2012 to April 2013, see **Table 1**. The reverse is observed for Bi where the highest and lowest concentration decreases in going from November 2012 to April 2013 (see **Table 1**), for example the highest concentration was 215.7 ppb in November 2012 while it decreases to 1.10 ppb in April 2013 (about 196 fold). For Sr and Ba, there are fluctuations in their concentration with significant variations. For Ba, the highest concentration in November 2012 is 165.9 while it is 58.5 ppb in March 2013. For Sr, the highest concentration was found to be 831.6 ppb in November 2012 while it is 623.6 ppb in March 2013.

4. CONCLUSION

High concentrations are observed for a number of elements for which no maximum allowed concentration values have been proposed (e.g., Li, Sr, Bi, Ba, V). These elements may warrant a toxicological assessment. These elements need urgent attention from a health perspective. Public authorities might be well advised to establish maximum allowed limits for such elements. This study demonstrates the necessity of documenting natural element concentrations and variation in drinking water resources on a regional scale. This must be carried out for as many elements as can be analyzed with today's techniques in different geological settings, to provide fingerprints of the different types of water.

5. ACKNOWLEDGEMENTS

The authors are grateful for the German research funding organisation (Deutsche Forschungsgemeinschaft), DFG for their financial support through TRION project.

REFERENCES

1. EU (European Union) Directive 98/83/EC on the Quality of Water intended for Human Consumption. OJ, L330, December 1998.
2. D. Banks, A. K. Midtgard, G. Morland, C. Reimann, T. Strand, K. Bjorvatn and U. Siewers, "Is Pure Groundwater Safe to Drink? Natural 'Contamination' of Groundwater in Norway," Geology Today, Vol. 14, No. 3, 1998, pp. 104-113.
3. B. T. Hart and T. Hines, "Trace Elements in Rivers," In: B. Salbu and E. Steinnes, Eds., Trace Elements in Natural Waters, CRC Press, Boca Raton, 1995, pp. 203-221.
4. C. Reimann, G. E. M. Hall, U. Siewers, K. Bjorvatn, G. Morland, H. Skarphagen and T. Strand, "Radon, Fluoride and 62 Elements as Determined by ICP-MS in 145 Norwegian Hard Rock Groundwater Samples," Science Total Environment, Vol. 192, No. 1, 1996, pp. 1-19.
5. C. Reimann, U. Siwers, H. Skarphagen and D. Banks "Influence of Filtration on Concentrations of 62 Elements Analysed on Crystalline Bedrock Groundwater by ICPMS," Science Total Environment, Vol. 234, No. 1-3, 1999, pp. 155-173.
6. S. Hesske, A. Parriaux and M. Bensimon, "Geochemistry of Spring Waters in Molasse Aquifers: Typical Mineral Trace Elements," Eclogae Geologicae Helvetiae, Vol. 90, No. 1, 1997, pp. 151-171.
7. A. Misund, B. Frengstad, U. Siewers and C. Reimann, "Natural Variation of 66 Elements in European Mineral Waters," Science Total Environment, Vol. 243, No. 244, 1999, pp. 21-41.
8. I. Kim and G. Kim, "Large Fluxes of Rare Earth Elements through Submarine Groundwater Discharge (SGD) from a Volcanic Island, Jeju, Korea," Marine Chemistry, Vol. 127, No. 1-4, 2011, pp. 12-19.
9. R. P. Janssen and W. Verweij, "Geochemistry of Some Rare Earth Elements in Groundwater, Vierlingsbeek, The

Netherlands," Water Research, Vol. 37, No. 6, 2003, pp. 1320-1350.

10. G. G. Gruau, A. Dia, G. Olivie-Lauquet, M. Davranche and G. Pinay, "Controls on the Distribution of Rare Earth Elements in Shallow Groundwaters," Water Research, Vol. 38, No. 16, 2004, pp. 3576-3586.

11. M. A. Rasheed, B. A. Radha, P. L. Srinivasa Rao, M. Lakshmi, J. Bala Chennaiah and A. M. Dayal, "Evaluation of Potable Groundwater Quality in Some Villages of Adilabad in Andhra Pradesh, India," Journal of Environmental Biolog, Vol. 33, 2012, pp. 689-693.

12. B. Frengstad, A. K. Midtgard, D. Banks, J. R. Krog and U. Siewers, "The Chemistry of Norwegian Groundwaters. III. The Distribution of Trace Elements in 476 Crystalline Bedrock Groundwaters as Analysed by ICP-MS Techniques," Science of the Total Environment, Vol. 246, No. 1, 2000, pp. 21-40.

13. C. Reimanna, K. Bjorvatnb, B. Frengstada, Z. Melakuc, R. Tekle-Haimanotc and U. Siewersd, "Drinking Water Quality in the Ethiopian Section of the East African Rift Valley I—Data and Health Aspects," The Science of the Total Environment, Vol. 311, No. 1-3, 2003, pp. 65-80.

14. "Environmental Profile for the West Bank, Volume 3: Hebron District," Applied Research Institute, Jerusalem, 1995.

15. A. Abed and S. Wishahi, "Geology of Palestine: The West Bank and Gaza Strip. Palestine," Palestine Hydrology Group, 1999.

16. F. M. Anayah and M. N. Almasri, "Trends and Occurrences of Nitrate in the Groundwater of the West Bank, Palestine," Applied Geography, Vol. 29, No. 4, 2009, pp. 588-601.

17. PHG, "Water For Life," Water, Sanitation and Hygiene Monitoring Program (WaSH MP), 2007/2008.

18. WHO, "Guidelines for Drinking-Water Quality, Addendum to Vol. 1. Recommendations," 2nd Edition, World Health Organisation, Geneva, 1998, pp. 10-11.

19. OMEE, "Monitoring Data for Uranium—1990-1995. Toronto, Ontario," Ontario Ministry of Environment and Energy, Ontario Drinking Water Surveillance Program, 1996.

20. P. Kurttio, A. Auvinen, L. Salonen, H. Saha, J. Pekkanen, I. Makeläinen, S. B. Vaisanen, I. M. Penttila and H. Komulainen, "Renal Effects of Uranium in Drinking Water," Environmental Health Perspectives, Vol. 110, No. 4, 2002, pp. 337-342.

21. A. C. Hakonson-Hayes, P. R. Fresquez and F. W. Whicker, "Assessing Potential Risks from Exposure to Natural Uranium in Well Water," Journal of Environmental Radioactivity, Vol. 59, No. 1, 2002, pp. 29-40.

22. J. Fitzgerald, D. Cunliffe, S. Rainow, S. Dodds, S. Hostetler and G. Jacobson, "Groundwater Quality and Environmental Health Implications, Anangu Pitjantjatjara Lands, South Australia. Canberra," Bureau of Rural Sciences, 2000.

23. WHO, WHO/SDE/WSH/03.04/118/Rev/1, Background Document for Development of WHO Guidelines for Drinking Water Quality, 2012.

24. S. Hirano and K. T. Suzuki, "Exposure, Metabolism, and Toxicity of Rare Earths and Related Compounds," Environ Health Perspective, Vol. 104, No. 1, 1996, pp. 85-95.

25. P. L. Smedley and D. G. Kinniburgh, "A Review of the Source, Behavior and Distribution of Arsenic in Natural Waters," Applied Geochemistry, Vol. 17, No. 5, 2002, pp. 517-568.

26. WHO, "Guidelines for Drinking Water Quality," World Health Organisation, Geneva, 1993.

27. US EPA, "Consumer Factsheet on: Nitrates/Nitrites," 2003. http://www.epa.gov/safewater/dwh/c-ioc/nitrates.html [Paper reference:2]

28. WHO, "Guidelines for Drinking Water Quality," 3rd Edition, Vol. 1, Recommendations, Geneva, 2004.

29. European Union, "Council Directive 98y83yEC of 3 November 1998 on the Quality of Water Intended for Human Consumption," Official Journal of the European Community, Vol. L330, 1998, pp. y32-y54.

30. BIS (Bureau of Indian Standards), "Specification for Drinking Water, Indian Standard (IS: 10500)," New Delhi, 1993.

31. T. H. Jaing, I. H. Hung, H. T. Chung, C. H. Lai, W. M. Liu and K. W. Chang, "Acute Hypermagnesemia: A Rare Complication of Antacid Administration after Bone Marrow Transplantation," Clinica Chimica Acta, Vol. 326, No. 1-2, 2002, pp. 201-203.

32. WHO, "Calcium and Magnesium in Drinking-Water: Public Health Significance," Library Cataloguing-in-Publication Data, Geneva, 2009.

33. W. M. Edmunds and P. L. Smedley, "Groundwater Geochemistry and Health. Overview," In: J. D. Appleton, R. Fuge and J. G. H. Mc. Call, Eds., Enviromental Geochemistry and Health with Special Reference to Developing Countries, Geological Society, Special Publication, Vol. 113, 1996, pp. 91-105.

34. US Environmental Protection Agency, "Strontium," 2012. http://www.epa.gov/rpdweb00/radionuclides/strontium.html

35. J. L. Seifter, "Potassium Disorders," In: L. Goldman, D. Ausiello and C. Medi, Eds., Cecil Medicine, 23rd Edition, Chapter 118, Saunders Elsevier, Philadelphia, 2007.

36. WHO, "Potassium in Drinking Water: Background Document for Development of WHO Guidelines for Drinking Water Quality," Geneva, 2009.

37. WHO, "Barium in Drinking-Water Background Document for Development of WHO, Guidelines for DrinkingWater Quality," Geneva, 2004.

5

CHAPTER

Modeling Preparative Chromatographic Separation of Heavy Rare Earth Elements and Optimization of Thulium Purification

Mark Max-Hansen, Hans-Kristian Knutson, Christian Jönsson, Marcus Degerman, Bernt Nilsson

Department of Chemical Engineering, Lund University, Lund, Sweden

ABSTRACT

Rare Earth Elements are in growing demand globally. This paper presents a case study of applied mathematical modeling and multi objective optimization to optimize the separation of heavy Rare Earth Elements, Terbium-Lutetium, by means of preparative solid phase extraction chromatography, which means that an extraction ligand, HDEHP, is immobilized on a C18 silica phase, and nitric acid is used as an eluent. An ICP-MS was used for online detection of the Rare Earths. A methodology for calibration and optimization is presented, and applied to an industrially relevant mixture. Results show that Thulium is produced at 99% purity, with a productivity of 0.2 - 0.5 kg Tm per m^3stationary phase and second, with Yields from 74% to 99%.

Keywords: Rare Earth Elements, Chromatography, Model Calibration, Optimization, Multi-Objective, HDEHP

1. INTRODUCTION

The Rare Earth Elements (REEs) are Scandium and Yttrium and the members of the Lanthanide group in the periodic table. Many of these metals are used in batteries, lasers, capacitors, superconductors [1] , which make the purification process demanding through high purity requirements in some of these applications. Current industrial separation methods include liquid-liquid extraction, selective oxidation/reduction, and ion exchange chromatography [2] -[5] . Extraction chromatography is commonly used for analytical purposes [1] [6] - [9] .

The REEs are of growing economic importance, with China limiting the export quotas [10] [11] , which creates a demand for alternative sources, and alternative purification processes, as the currently dominating liquid- liquid extraction uses high quantities of organic compounds for the separation. Solid phase extraction or extraction chromatography uses smaller amounts of organic chemicals than liquid-liquid extraction. This is why solid phase extraction could be one of the alternative purification processes.

This work is a continuation of the work presented in [12] - [14] where a mixture of the elements from Neodymium to Gadolinium was used in modeling and optimization of a chromatographic step, and also in experimental optimization. In the present work a mixture of all REEs except Scandium was separated on a modified column that operates in a solid phase extraction mode, meaning that the extractant is immobilized as a ligand in the column. To find the optimal operating conditions for the separation, a Dispersive-Equilibrium model was constructed, based on Langmuirian type binding with mobile phase modulator. The parameters for the model were calibrated from isocratic and gradient experiments. The optimal operating points were found by applying an optimization algorithm on the calibrated model and performance functions. In addition to this one of optimal operating points was verified experimentally.

2. THEORY

To design and optimize the chromatographic system, a performance function is required. This in turn depends on a mathematical formulation of the physical chromatography system. This mathematical formulation requires retention parameters, mass transfer parameters and column capacity to be calibrated. To reduce the complexity of the calibration, only the retention parameters are calibrated with a less computationally expensive model, i.e. the retention volume model. These retention parameters are then used as a starting guess in the more advanced model of the chromatography column, which takes dispersion and mass transfer effects into account as well as the retention data. For the optimization, Productivity and Yield were selected as performance functions, subject to a Purity constraint.

2.1. The Retention Volume Model

The retention volume model is an ideal model, which can be seen as a simplified version of the film mass transfer model, presented in 2.2, that includes the isocratic modifier concentration effect on the retention volume:

$$V_R = V_{R0} + (1-\varepsilon_c) \cdot \varepsilon_P \cdot K_D \cdot H \tag{1}$$

where V_R is the retention volume, V_{R0} is the retention volume under non-binding conditions, ε_C is the column porosity, ε_P is the particle porosity, K_D is the exclusion factor and H is the Henry constant [15] :

$$H = H_0 \cdot s_{\text{modifier}}^{-\nu} \tag{2}$$

where H_0 is the Henry constant when the modifier concentration is unity, s_{modifier} is the modifier concentration (mol/m^3), ν is the stochiometric coefficient. This expression is valid only when the loaded amount is much smaller than the total adsorption capacity.

2.2. The Dispersive-Equilibrium Model

The model including mass transfer and non-linear effects is a Dispersive-Equilibrium model using Langmuir type binding with mobile phase modifier, consisting of a dispersive-equilibrium description of the mobile phase with all mass transfer effects lumped in the dispersion, and Langmuir equilibrium adsorption for the solid phase [16] . For component i, its liquid-phase concentration in the column is described by the following equation:

$$\frac{\partial c_i}{\partial t} = D_{AX} \cdot \frac{\partial^2 c_i}{\partial x^2} - v_{app} \frac{\partial c_i}{\partial x} - \frac{1}{\varepsilon_T} \cdot \frac{\partial q_i}{\partial t} \tag{3}$$

where c_i is the concentration of component i in the mobile phase (mol/m^3), D_{AX} is the apparent dispersion coefficient (m^2/s) see Equations (6)-(8). x is the axial coordinate along the column (m), v_{app} is the apparent velocity, which is the superficial velocity divided by the total void fraction (m/s), ε_T is the void fraction of the packed bed (m^3 mobile phase/m^3 column), q_i is the concentration of adsorbed species i (mol/m^3 column) and t is the time (s).

The column equation is subject to two boundary conditions, one at the inlet and one at the outlet. A Robin boundary condition describes the concentration at the inlet:

$$\left. \frac{\partial c_i}{\partial x} = \frac{v_{app}}{D_{AX}} \left(c_i - c_{inlet,i} \right) \right|_{x=0} \tag{4}$$

where $c_{inlet,i}$ is the concentration of component i at the inlet. The concentration of species i at x = 0 may be lower than the inlet concentration due to dispersion. The other boundary condition is at the outlet of the column, at x = L, and is a von Neumann condition:

$$\left. \frac{\partial c_i}{\partial x} = 0 \right|_{x=L} \tag{5}$$

The concentration of species i adsorbed on the stationary phase is described by the following equation, which is a mobile phase modified version of the Langmuir adsorption description [21] :

$$0 = H_0 \cdot s^{-\nu} \cdot c_i \cdot \left(1 - \sum \frac{q_i}{q_{\max,i}}\right) - q_i \tag{6}$$

where H_0 is the Henry constant, s is the modifier concentration, ν is the stochiometric constant, $q_{max,i}$ is the maximum adsorption capacity for component i. This equation must at equilibrium satisfy the adsorption equilibrium isotherm, which means that the parenthesis must be 0 at equilibrium.

The flow rate dependence of the apparent dispersion, including mass transfer resistance and kinetics, is approximated with a second degree polynomial. This comes from the Knox equation [17] :

$$h = \frac{B}{v_{app}/d_p} + Av_{app}^{\frac{1}{3}}/d_p + Cv_{app}/d_p \tag{7}$$

where h is the reduced plate height, v_{app} is the apparent velocity, d_p is the particle diameter. A, B and C are constants coming from flow structure and non equilibrium effects. The apparent axial dispersion can then be described as:

$$D_{AX} = \frac{h}{2v_{app}} \tag{8}$$

where D_{AX} is the apparent dispersion coefficient. Combining equations 7 and 8 give:

$$D_{AX} = \frac{1}{2} \cdot \left(B + Av_{app}^{\frac{4}{3}} + Cv_{app}^2\right) \tag{9}$$

Estimating the apparent dispersion with a second degree polynomial gives a small error in the second term, however this is acceptable with current model precision.

2.3. Optimization

The target component for the optimization is Thulium, as producing pure Thulium would produce pure Ytterbium also, and then both of the major components have been purified, see Figure 4 andFigure 5. The rest of the REEs are considered impurities. The optimal performance of the separation system depends on the technical and economic constraints and incentives that are placed upon this system. Common performance attributes used are Productivity, Yield, Purity and Specific Productivity. In this work, Productivity, presented in Equation (10) and Yield, presented in Equation (11) are used as performance functions, while the Purity is used as a constraint. The productivity can be calculated from:

$$Pr = \frac{\int_{t_{cut,1}}^{t_{cut,2}} F \cdot c_{outlet} \, \mathrm{d}t}{t_{cycle} \cdot V_{col}} \tag{10}$$

where $t_{cut,1}$ and $t_{cut,2}$ are the times between which the target component can be collected at required purity (s), F is the flow rate (m^3/s), c_{outlet} is the outlet concentration of the component (mol/m^3), t_{cycle} is the total cycle time (s) and V_{col} is the column volume (m^3). The yield can be described as:

$$Y = \frac{\int_{t_{cut,1}}^{t_{cut,2}} F \cdot c_{outlet} \, \mathrm{d}t}{\int_{0}^{t_{load}} F \cdot c_{feed} \, \mathrm{d}t} \tag{11}$$

In this case t_{load} is the time at the end of the loading(s), c_{feed} is the concentration in the feed (mol/m^3).

When two competing objectives, in this case productivity and yield are competing, the result will be a Pareto front, which means that for a given optimal point on the front, increasing one objective must decrease the other [18] - [20] .

3. MATERIALS AND METHODS

3.1. Materials

The experiments were performed on an Agilent 1260 Bio-Inert HPLC system, consisting of two Agilent 1260 Bio-Inert Quaternary pumps, one Agilent Bio-Inert 1260 Autosampler and one Agilent 1260 Bio-Inert UV/Vis detector connected to an Agilent 7700 ICP-MS for detection of the rare earths. The column preparation was performed using a GE Healthcare ÄKTA Purifier 100. The individual rare earths were acquired from Merck at 1 g/L in nitric acid. The concentrated nitric acid of HPLC grade was acquired from Merck. The columns used were Kromasil C18, 100 Å, 16µm columns with immobilized HDEHP acting as ligand. HDEHP was acquired from Merck. Methanol was acquired from Merck. To be able to do higher flow rate experiments, an extra isocratic pump was added to the system after the column, and used to feed the ICP-MS with a constant flow of 1 ml/min, the rest going to waste or collection. The heavy mixture of rare earths used in the Thulium optimization was mixed to mimic an industrially relevant REE-stream, with a composition as described in Table 1.

Column Preparation

The column was modified by passing a mixture of 22 mM HDEHP in 65% (v/v) Methanol-water, using the conductivity meter of the ÄKTA to determine the breakthrough volume, and thereby the ligand concentration of 345 mol/m^3 column volume. As the ligand forms dimers, the concentration of sites was 172.5 mol/m^3 column volume.

3.2. Methods

Isocratic experiments at 16 compositions of the mobile phase were performed to get retention volumes for all components, gradient experiments at different flow rates were performed to get peak broadening from mass transfer and kinetics.

Table 1. Composition of heavy mixture, 336 g/L, 14 M HNO_3.

	% (wt)
Y	0.67
La	1.2
Ce	0.1
Pr	0.09
Nd	0.13
Sm	0.42
Eu	0.06
Gd	1.59
Tb	0.13
Dy	0.74
Ho	4.23
Er	6.54
Tm	11.99
Yb	71.89
Lu	0.24

3.2.1. Experimental Design for Calibration

16 isocratic experiments, 0.35 to 4.9 M nitric acid were performed, with 3 replicates on all experimental points. The experiments were performed at 0.35, 0.53, 0.56, 0.60, 0.63, 0.67, 0.70, 1.05, 1.40, 1.75, 2.10, 2.8, 3.50, 4.20 and 4.90 M nitric acid concentration, using the quaternary pump to create the isocratic eluent.

No overloading experiments were performed, as the column capacity had been decided during the preparation of the column.

3.2.2. Chromatography Method

The chromatographic experiments performed, consisted of four consecutive steps; loading, isocratic elution, Cleaning in Place (CIP), and reequilibration.

The injected amount during loading was 10 μl of a mixture containing 0.0357 g/L of each of the lanthanides, plus yttrium. This was well within the linear range of the isotherm, so any competitive effects could be neg- lected. After loading, a prepared eluent of a predetermined concentration was passed through the column for 120 minutes, this was due to a limitation of the ICP-MS software, after elution a CIP at 7M nitric acid was performed. After the CIP the system was reequilibrated with water. All experiments were performed with three replicates.

3.2.3. Validation Experiment Chromatography Method

For the validation experiment, the experimental setup was modified so that it no longer had the post column pump. This decreased the post column dispersion. The validation experiment was adjusted somewhat from the very aggressive points on the Pareto front, to a longer gradient and with a lower flow rate. The validation experiment consisted of five consecutive steps: loading, wash, gradient elution, CIP and reequilibration. The injected amount during loading was 200 μl heavy mixture, diluted to 33.6 g/L with deionized water, then a 4.5 ml wash with deionized water, followed by a 17 ml gradient from 0 to 7 M nitric acid. Then 4 ml CIP at 7 M nitric acid, followed by 10 ml reequilibration with deionized water.

3.2.4. Simulation Method

The chromatography model was discretized in the axial direction, to turn the partial differential equation into a system of Ordinary Differential Equations (ODE) and Algebraic equations, Differential Algebraic Equation (DAE). The discretization used a 2 point backward finite difference scheme for the flow, and a 3 point central finite difference scheme for the dispersion. The assembled DAE system was solved with MATLAB ODE/DAE solver ode 15 s, which solves stiff DAEs. Discretization and Simulation is handled by the Preparative Chromatography Simulator, PCS, developed at the Department of Chemical Engineering, Lund University [21] . In order to reduce the size of the computational problem for the optimization, all components lighter than Gadolinium were lumped together and added to the mass of Gadolinium, this is done, as in the worst case perspective of separating Thulium from the heavy mixture, they would act as Gadolinium, where they would pass more readily through the column in reality.

3.2.5. Calibration Method

First the Henry constant and the stochiometric coefficient were calculated from the slope and intercept of the logarithm of the modified retention volume, versus the logarithm of the modifier concentration. These were then used for the inverse method, i.e. minimizing the least squares error for the normalized detector response and the normalized simulated chromatograms, where the mass transfer effects were fitted, using a Nelder-Mead simplex algorithm. The maximum binding capacity of the columns was decided during preparation of the columns, and based on the HDEHP concentration in the backbone.

3.2.6. Optimization Method

The calibrated model was used with the performance functions presented in 2.3, and used with a modified Differential Evolution (DE) algorithm and parallel computations in a cluster environment [13] . For the optimization, four decision variables were chosen, load volume, gradient length, CIP length and flow rate.

4. RESULTS

The modified columns had some known characteristics, such as the porosities, $\varepsilon_c = 0.6$, $\varepsilon_p = 0.4$, and the maximum binding capacities, which were decided during the impregnation of the column, and calculated from equation 12:

$$q_{\max,i} = \frac{\lambda}{\nu_i} \tag{12}$$

where λ is 172.5 (mol/m^3 column), and ν_i is the stochiometric constant for component i.

4.1. Model Calibration

The Henry constants and characteristic charges, presented in Table 2, were calculated from the retention volume experiments presented in Figure 1. There is a wide span in retention data for the replicates, as can be seen in Figure 1. This is probably a result of non-isocratic and inexact mixing of the eluent as a result of the rotating inlet valve combined with the non-ideal mixing properties of water and nitric acid.

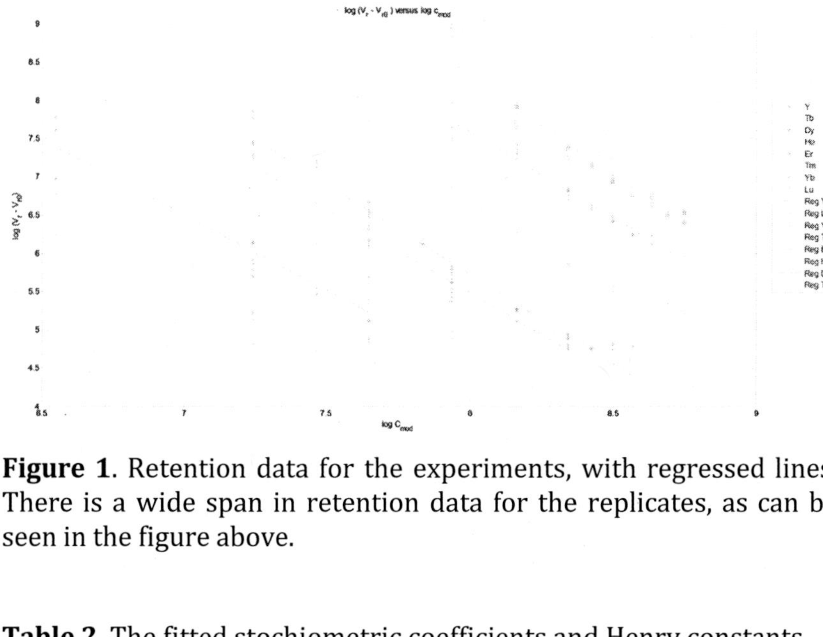

Figure 1. Retention data for the experiments, with regressed lines. There is a wide span in retention data for the replicates, as can be seen in the figure above.

Table 2. The fitted stochiometric coefficients and Henry constants.

Component	v	H	q_{max}
Yttrium	2.18	22.96	79
Terbium	2.20	21.07	78
Dysprosium	1.96	20.24	88
Holmium	2.12	21.89	81
Erbium	2.34	24.49	73
Thulium	2.19	24.29	78
Ytterbium	2.25	25.60	77
Lutetium	2.64	29.47	65

4.2. The Flow Rate Experiments

To calibrate the mass transfer effects, flow rate experiments were performed at 1, 1.25, 1.5, 1.75, 2, 2.5 and 3 ml/min, the normalized detector response was used against a normalized simulated response from the PCS, and parameters fitted by a least squares approach. The experimental and model results are presented in Figure 2, with model parameters for the axial dispersion as in Equation (9), with the terms: A = 1.05 ´ 10^{-8}, B = 1.00 ´ 10^{-8} and C = 0.9936 ´ 10^{-8}.

4.3. Optimization

The multi-objective optimization was performed using Productivity and Yield, see Equations (10) and (11), as objective functions. The optimization output is a the Pareto front, presented in Figure 3, which gives the system performance for different operating points, comparing in this case Productivity and Yield, under a 99% Purity constraint for the Thulium. Figure 3 shows that the Pareto front is much denser in the high yield region, than it is in the high productivity region. This is to be expected, as high yields are relatively easy to achieve by keeping the loading volume small, and the gradient slope flat. The selected points in Figure 3 are presented in Figure 4, with their decision variables presented in Table 3.

Figure 2. Model response and experimental response for dysprosium at different flow rates. There is a reasonably good fit, and using a least squares approach to model calibration will result in wider peaks where there is variation in the experiments.

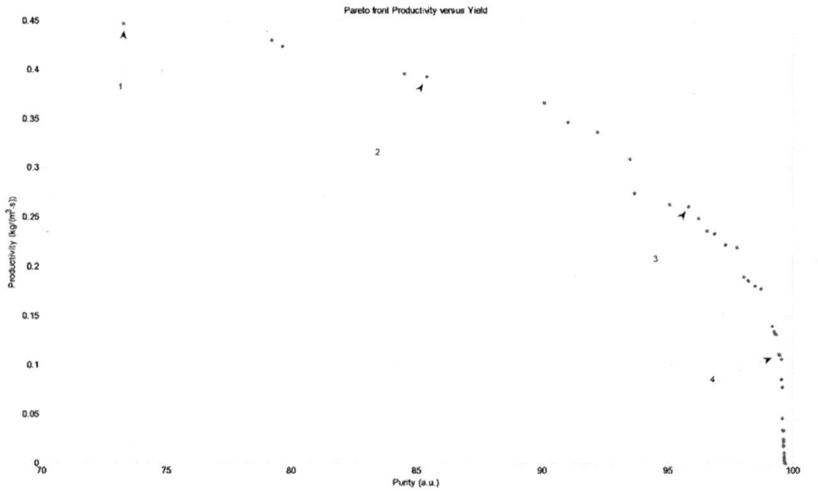

Figure 3. Pareto front Productivity versus Yield, selected Pareto points correspond to chromatograms in Figure 4.

All of the operating points are under overloaded conditions, as can be seen in Figure 4, with a decreasing degree of overloading as we move from high productivity in chromatogram 1 to high yield in chromatogram 4. The curvature of the Pareto front is a result of peak overlapping, so that in chromatogram 4, the Thulium peak is slightly contaminated on the back side by Ytterbium but base line separated at the front, then going to chromatogram three, where the Thulium peak is contaminated more by Ytterbium at the back side, and starting to see Erbium at the front side. Chromatogram 2 shows even more Ytterbium in the back side of the peak, and a bit of Erbium at the front side as well. In chromatogram 1, the Erbium and Ytterbium peaks almost meet underneath the Thulium peak.

4.4. Validation

As described in 3.2.2, the validation experiment was performed at a slightly less aggressive operating point than max (Prod * Yield), with a longer gradient and a flow rate of 1 ml/min. The chromatogram from the validation experiment, presented in Figure 5 shows that base line separation on the front side of the Thulium peak is possible, with a small overlap at the back side, while both Thulium and Ytterbium show overloaded conditions. This chromatogram has a calculated Productivity of 0.20 kg Thulium per cubic meter of column volume and second, with a Yield above 95%. The reason that the peaks are a lot sharper in the validation chromatogram than the simulated peaks, is that the post column pump was removed, which leads to much less post column dispersion.

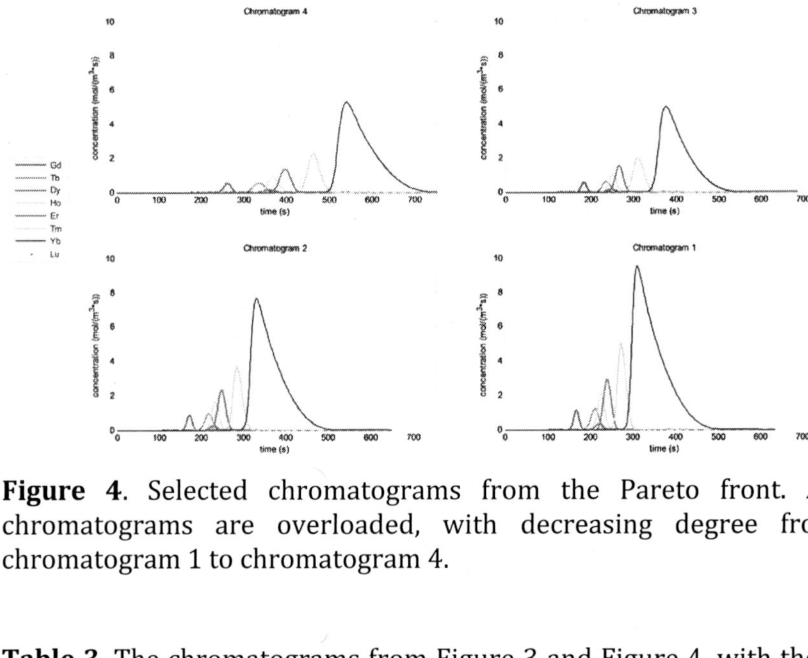

Figure 4. Selected chromatograms from the Pareto front. All chromatograms are overloaded, with decreasing degree from chromatogram 1 to chromatogram 4.

Table 3. The chromatograms from Figure 3 and Figure 4, with their decision variable values and Productivity and Yield values.

	Prod (kg/m^3·s)	Yield (%wt)	Load (ml)	Gradient length (ml)	Flow rate (ml/min)
Chromatogram 1	0.181	98.5	0.0684	7.70	1.26
Chromatogram 2	0.263	95.1	0.0672	5.17	1.55
Chromatogram 3	0.396	84.5	0.110	5.62	1.68
Chromatogram 4	0.447	73.3	0.140	5.66	1.73

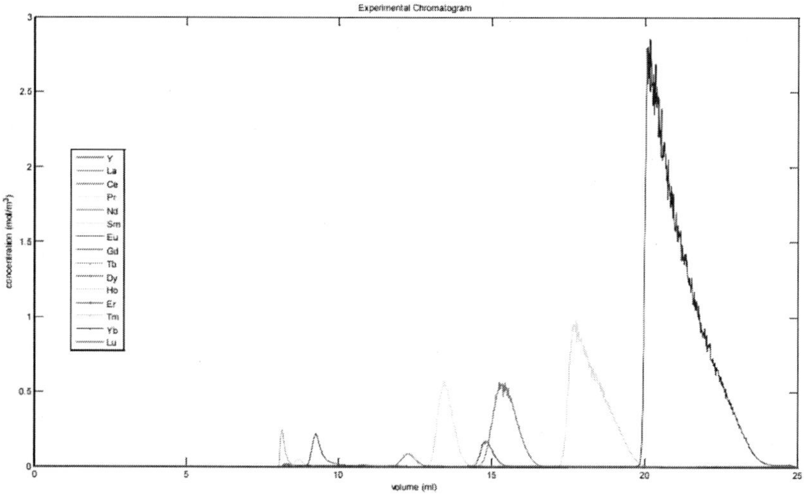

Figure 5. Experimental chromatogram. This slightly overloaded chromatogram shows base line separation at the front side of the Thulium peak, and a slight overlap on the back side of the peak. This means high purity and high yield.

5. CONCLUSION

Mark Max-Hansen, Hans-Kristian Knutson, Christian Jönsson, Marcus Degerman, Bernt Nilsson (m^3·s) to 0.45 kg/(m^3·s) at Yields of 98.5% to 73.3%.

REFERENCES

1. Verma, S.P. and Santoyo, E. (2007) High-Performance Liquid and Ion Chromatography: Separation and Quantification Analytical Techniques for Rare Earth Elements. Geostandards and Geoanalytical Research, 31, 161-184.

2. Pierce, T.B. and Hobbs, R.S. (1963) The Separation of the Rare Earths by Partition Chromatography with Reversed Phases: Part 1. Behavior of Column Material. J Chrom A, 12, 74-80.

3. Russell, R.G. and Pearce, D.W. (1943) Fractionation of the Rare Earths by Zeolite Action. Journal of the American Chemical Society, 65, 595-600.

4. Lister, B.A. and Smith, M.L. (1948) The Fractionation of Cerium (III) and Neodymium Mixtures on an Ion-Exchange Column. Journal of the Chemical Society, 1272-1275.

5. Choppin, G.R. and Silva, R.J. (1956) Separation of the Lanthanides by Ion Exchange with Alpha-Hydroxy Isobutyric Acid. Journal of Inorganic and Nuclear Chemistry, 3, 153-154.

6. Harris, D.H. and Tompkins, E.R. (1947) Ion-Exchange as a Separation Method; Separations of Several Rare Earths of the Cerium Group (La, Ce, Pr and Nd). Journal of the American Chemical Society, 69, 2792-2800.

7. Kettele, B.H. and Boyd, G.E. (1947) The Exchange Adsorption of Ions from Aqueous Solutions by Organic Zeolites. IV. The Separation of the Yttrium Group Rare Earths. Journal of the American Chemical Society, 69, 2800-2812.

8. Spedding, F.H., Powell, J.E. and Wheelwright, E.J. (1956) The Stability of the Rare Earth Complexes with N-Hydroxy- ethyl Ethylenediaminetriacetic Acid. Journal of the American Chemical Society, 78, 34-37.

9. Minczewski, J. and Dybczynski, R. (1961) Separation of Rare Earths on Anionexchange Resins. III. The Position of Scandium in the Separation Scheme of Rare Earth Complexes with Ethylenediaminetetraacetic Acid. Journal of Chromatography A, 7, 79-96.

10. Meyer, L. and Bras, B. (2011) Rare Earth Metal Recycling. 2011 IEEE International Symposium on Sustainable Systems and Technology (ISSST), Chicago, 16-18 May 2011, 1-6.

11. Jordens, A., Cheng, Y.P. and Waters, K.E. (2013) A Review of the Benefication of Rare Earth Element Bearing Minerals. Minerals Engineering, 41, 97-114.

12. Ojala, F., et al. (2012) Modelling and Optimisation of Preparative Chromatographic Purification of Europium. Journal of Chromatography A, 1220, 21-25.

13. Max-Hansen, M., et al. (2011) Optimization of Preparative Chromatographic Separation of Multiple Rare Earth Elements. Journal of Chromatography A, 1218, 9155-9161.

14. Knutson, H. (2014) Experimental Productivity Rate Optimization of Rare Earth Element Separation through Preparative Solid Phase Extraction Chromatography. Journal of Chromatography A, 1348, 47-51.

15. Karlsson, D. (2004) Aspects of the Design of Preparative Chromatography for Purification of Antibodies. Ph.D. Thesis, Lund University, Lund.

16. Jakobsson, N., et al. (2005) Optimisation and Robustness Analysis of a Hydrophobic Interaction Chromatography Step. Journal of Chromatography A, 1099, 157-166.

17. Guichon, G., Felinger, A., Shirazi, D.G. and Katti, A.M. (2006) Fundamentals of Preparative and Nonlinear Chromatography. 2nd Edition, Elsevier Inc., San Diego.

18. Price, K., Storn, R. and Lampinen, J. (2005) Differential Evolution: A Practical Approach to Global Optimization. Springer, Berlin. [Paper reference:1]

19. Wang, Y.-N., et al. (2010) Multi-Objective Self-Adaptive Differential Evolution with Elitist Archive and Crowding Entropy-Based Diversity Measure. Soft Computing, 14, 193-209.

20. Holmqvist, A., et al. (2013) A Model-Based Methodology for the Analysis and Design of Atomic Layer Deposition Processes—Part III: Constrained Multi-Objective Optimization. Chemical Engineering Science, 96, 71-86.

21. Borg, N. (2013) Modeling and Calibration of Preparative Chromatography. Ph.D. Thesis, Lund University, Lund.

6

CHAPTER

Characterization and Iron Removal Treatment of Ion-Adsorption Rare Earth Tailings in Southern China

Yongqing Wang, Xin Nie, Qibing Chang*, Huayin Liang, Xiaozhen Zhang, Jian-Er Zhou

School of Materials Science and Engineering, Jingdezhen Ceramic Institute, Jingdezhen, China

ABSTRACT

The ion-adsorption rare earth tailings have become a serious environmental pollution in Southern China, yet the potential of their economical value has not been fully exploited. In this work, the chemical and mineral compositions of the ion-adsorption rare earth tailings were characterized by Mineral Liberation Analyze (MLA) and XRF. The results show that 91.98 wt% of the tailings are composed of kaolinite and quartz, latter of which was removed by the sieving method. The other minor minerals contain feldspar, biotite, muscovite, titanomagnetite and limonite. Amongst these, the iron-bearing minerals are mostly found in the titanomagnetite and limonite which can be mostly removed by using a periodic high-gradient magnetic separator with a magnetic induction of 0.6 Tesla. The Fe_2O_3 content of the tailings changed from 2.11 wt% to 1.06 wt% after the sorting process, which met the Chinese national standard of TC-3 grade raw materials for ceramic industry applications. The Fe_2O_3 content in kaolinite was further decreased after $Na_2S_2O_4$treatment.

Keywords: Tailings, Resource, Rare Earth, Iron Removal, Kaolinite, Magnetic Separation

1. INTRODUCTION

Rare earth elements (REEs) have become vital and indispensable components of many advanced products, devices and technologies. The production of rare earth minerals in China increased by an average of 40% annually between 1978 and 1989, and had provided more than 90% of the world's supply since 2001 [1] - [3] . Ion-adsorp- tion rare earth deposition was firstly discovered in Ganzhou, Southern China, in 1970. Unlike other rare earth minerals which are in solid state mineral phase and tend to be associated with the radioactive elements uranium and thorium, ion-adsorption rare earth minerals occur at a simple trivalent cationic state, which is simply adsorbed onto clays and could be readily extracted by a simple leaching technique with an aqueous electrolyte solution via anion-exchange process [4] . The extraction of ion-adsorption rare earth is carried out by surface/ mountain-top mining followed by tank or heap leaching with sodium chloride ammonium sulfate solution.

Ion-adsorption rare earth deposition has a relatively low REEs concentration ranging from 10 to a few hundred parts per million by weight [4] . Using heap leaching techniques, it is estimated that for the production of 1 tons (t) rare earth oxide, 2000 tons tailings are disposed into adjacent valleys and streams. Due to this, the rare earth miners in Ganzhou region have generated 191 million tons of tailings [5] . These hazardous tailings are a major environmental concern causing increased water and soil loss, deterioration of water quality and potentially other geological disasters. Therefore, the re-utilization of the tailings not only reduces their impact of environmental pollution but also can be used to develop highly valued end-products for the associated industries [3] . To realize it, the tailings should be characterized and pre-treated before their re-utilization.

In the present work, the chemical and mineral compositions of the ion-adsorbed rare earth tailings from the Southern China were analyzed in order to determine the feasibility of the tailings as raw materials for the ceramic industry. The treatments of the tailings, such as sieving, iron removal treatment, were also considered.

2. MATERIALS AND METHODS

The tailings were collected from the ion-adsorption rare earths area in Ganzhou, Southern China. Representative samples were prepared by following Chinese standard method of mixing (China Standard Specification, GB2007.1-87).

The chemical element analysis of samples was performed with an XRF spectrometer (PANalytical Axios- mAX, Netherland).

The bulk mineral composition was performed with Mineral Liberation Analyzer (MLA 650, Czech Republic), which was equipped with environment SEM (FEI, Quanta 650), EDS (Bruker Quantax 200) and mineral parameters automatic analysis software of (MLA2.9). For preparation of grain mounts in epoxy blocks, 10 - 100 g of dried mineral sands is required. Aliquots of 10 - 20 g of a sample will be mixed with pure graphite powder to prevent particle cohesion and embedded in epoxy blocks. The preferred diameter of the round blocks is 30 mm, diameters of 25 mm. The other operational parameters are reported in Reference [6] .

In the pre-treatment processes, the tailings was sieved sequentially by the copper standard sieves with the sieve sizes of 0.8 mm, 0.4 mm, 0.25 mm, 0.12 mm, 0.075 mm, 0.043 mm. The −0.043 mm fraction of the tailings was used for the process of iron removal. Firstly, the sample was fed to the high gradient magnetic separator (SSS-I-145) [7] [8] to remove magnetite and ilmenite, followed by a bleaching process of the remained samples (fine kaolinite) using hydrogen peroxide [9] or sodium dithionite [10] as the leaching agent.

3. RESULTS AND DISCUSSION

3.1. Characterization of the Tailings

3.1.1. Chemical Composition of Tailings Characterized by XRF

The chemical composition of the tailings sample was characterized by XRF. The data is presented inTable 1. The rich chemical composition in oxides in mine tailing are SiO_2 (78.32 wt%), Al_2O_3 (17.50 wt%) and Fe_2O_3 (2.11 wt%). However, Fe_2O_3 is not what we want because it gives a brown color to ceramics. Therefore, iron removal is required to the tailings until the Fe_2O_3 content is under the maximum limit (1.8 wt%) for ceramic applications according to the Chinese standard GB/T 14563-1993.

3.1.2. Mineral Composition of Tailings Characterized by MLA

The tailings are the mixtures of many minerals. Generally, the minerals can be distinguished out by Mineral Liberation Analyzer (MLA). Table 2 shows the mineral composition of the tailings. Quartz and kaolinite are the major minerals of the tailings, which take up 91.98 wt% of the tailings. Quartz and kaolinite are commonly utilized in ceramic application. Therefore, they can be of industrial interest after a proper treatment processing. The other silicates, such as feldspar, biotite, muscovite and amphibole, are also generally used as sintering aid in ce ramic applications. The iron- and titanium-contained minerals include titanomagnetite, limonite, pyrite, ilmenite, rutile and leucoxene, which make up 1.72 wt% of the tailings. The total content of rare earth minerals is only 0.04 wt%, which include monazite, xenotime, parisite and colloidal rare earth.

The SEM images of the minerals are shown in Figures 1-5.

Table 1. Chemical composition of the tailings measured by XRF.

Composition	SiO_2	Al_2O_3	Fe_2O_3	TiO_2	K_2O	Na_2O	CaO	MgO
Content (wt%)	78.32	17.50	2.11	0.38	1.21	0.43	0.22	1.33

Table 2. Mineral composition of the tailings.

Mineral	Content wt%	Mineral	Content wt%	Mineral	Content wt%
Kaolinite	42.29	Monazite	0.01	Colloidal rare earth	0.01
Quartz	49.69	Xenotime	0.01	Titanomagnetite	1.10
Feldspar	3.90	Parisite	0.01	Limonite	0.29
Diopside	0.01	Rutile	0.09	Ilmenite	0.13
Muscovite	0.68	Leucoxene	0.10	Iron titanite	0.04
Biotite	1.36	Pyrite	0.01	Remainder	0.14
Amphibole	0.11	Galena	0.01	Total	100

Figure 1 shows the SEM images of kaolinite particles in the tailings. The kaolinite particles are observed as aggregation of microcrystalline flakes, which is consistent with the results in these references [9] [11] -[13] . The kaolinite grains are irregular in shapes. The particle size of the kaolinite grains is estimated in the range of 50 - 300 μm. It is also found that monazite, biotite and magnetite are embedded in the kaolinite grains. The chemical compositions and their concentrations of kaolinite measured by EDS are Fe_2O_3 0.57%, TiO_2 0.10%, Al_2O_3 39.24%, SiO_2 58.36%, K_2O 0.12%, CaO 0.19% and MgO 1.05% (in weight percent). Specially, it has been reported that the magnetite grains embedded in the kaolinite grains may not be removed by magnetic method [8] or chemical method [9] [10] . Therefore, 0.24 wt% Fe_2O_3 of the tailings (0.57% * 0.42 (kaolinite content in tailings) = 0.24%) still remain in the tailings even if after iron removal.

The chemical compositions and their concentrations of feldspar measured by EDS are K_2O 7.50%, Na_2O 4.62%, CaO 0.51%, Al_2O_3 19.59%, SiO_2 67.58% and FeO 0.20% (in weight percent). Fe element is incorporated in the crystal lattice of feldspar. Most of feldspar has been weathered into kaolinite.

Quartz is the major mineral in the tailings according to the data in Table 2. Figure 2 shows the SEM images of the quartz particles. As can be seen, the quartz is associated with magnetite, muscovite, or parisite.

Magnetite is the main iron-bearing minerals in the tailings. EDS results show the chemical elements of the magnetite mineral include

Fe 66.28%, Ti 3.18%, Mn 0.51%, Zn 0.49%, K 0.01%, Al 0.95%, Si 0.67% and O 27.92% (in weight percent).

The SEM image of magnetite particles associated with the TiO_2 is shown in Figure 3(A). The titanomagnetite mineral forms a grid mark pattern in the cross-section of magnetite grain. Figure 3shows that it is hard to separate titanomagnetite from magnetite by sieving separation method.Figure 3(B) shows the image of the magnetite associated kaolinite. In fact, Figure 1(D) and Figure 3(B) show that kaolinite and magnetite have the similar associated mineral structure. This associated structure results in the difficulty in the separation of magnetite from kaolinite.

Limonite is another iron-bearing mineral found in the tailings. EDS results show the chemical elements of the magnetite mineral include Fe 53.19%, Ti 0.52%, Mn 0.62%, Zn 0.08%, Ca 0.24%, Al 3.57%, Si 3.05%, Cr 0.11%, S 0.44% and O 38.12% (in weight percent).

Figure 4 shows the SEM image of limonite particles. Limonite particles found to be are granular and dust- like. The particle size is in the range of 0.1 μm - 0.08 mm. 80 wt% of limonite has particle sizes of less than 0.01 mm. And most of the dust-like limonite particles are mixed with kaolinite.

The chemical composition of ilmenite in the sample is 52.38% TiO_2, 41.23% FeO and 6.40% MnO (in weight percent). Ilmenite in the tailings is granular and very fine. The particles size distribution is in the range of 0.1 μm - 0.15 mm. Ilmenite grains are mostly associated with magnetite, kaolinite or xenotimeetc, as shown in Figure 5.

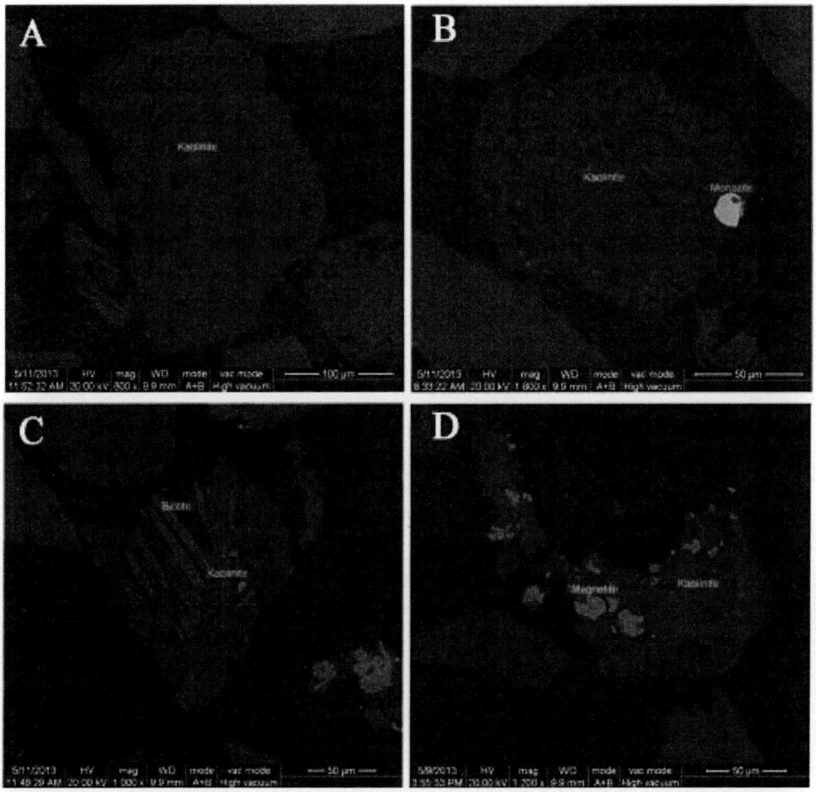

Figure 1. SEM images of kaolinite: (A) single grains; (B) embedded monazite; (C) embedded biotite; (D) embedded magnetite.

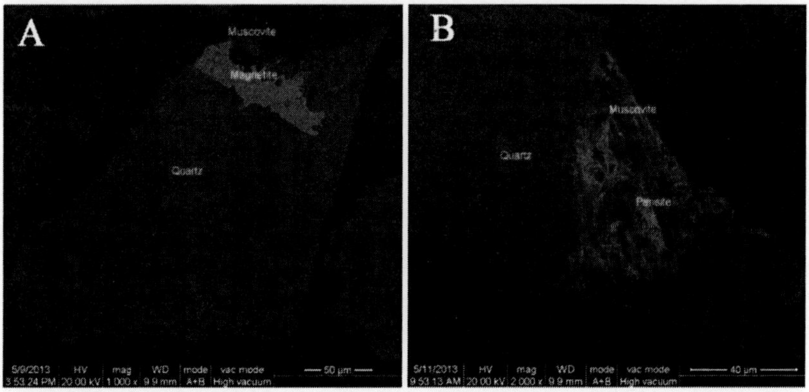

Figure 2. SEM images of the associated minerals of (A) quartz, magnetite and muscovite; (B) quartz, muscovite and parasite.

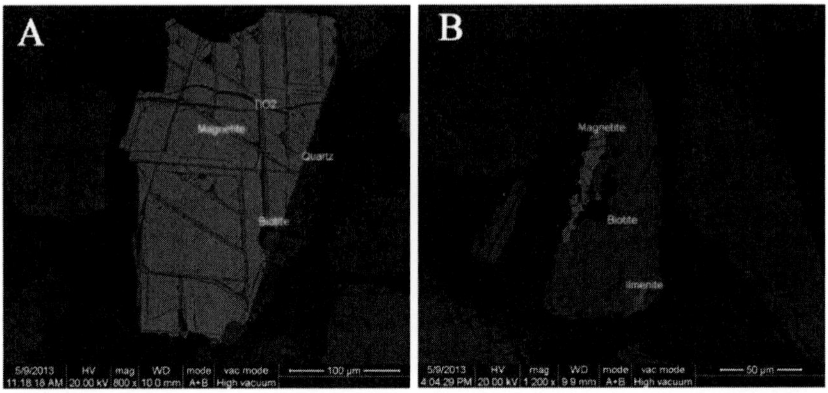

Figure 3. SEM images of the associated minerals of (A) magnetite and TiO_2; (B) magnetite and kaolinite.

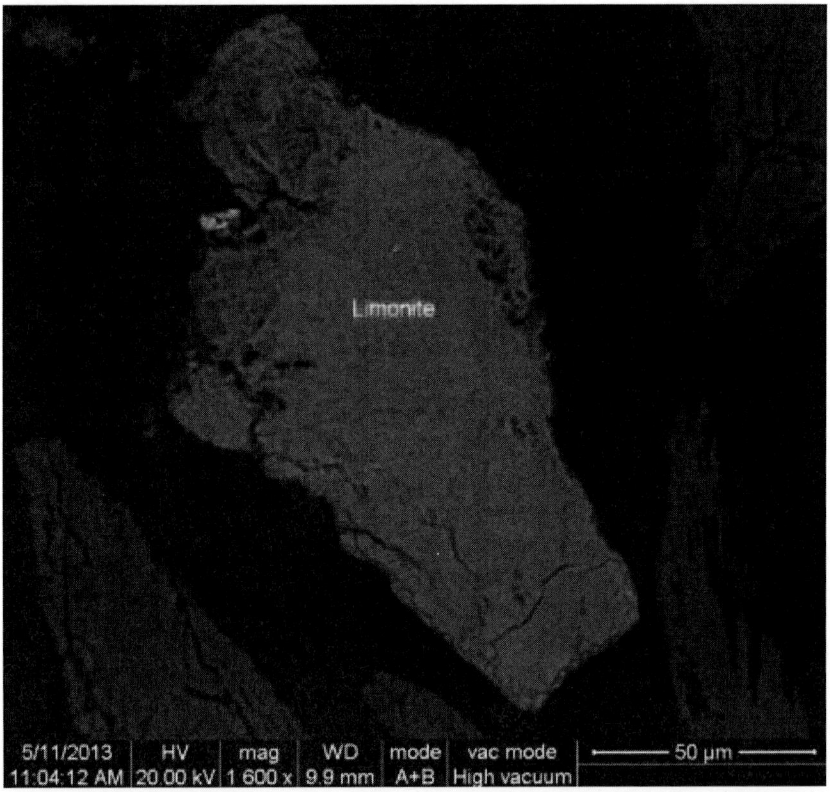

Figure 4. SEM image of limonite.

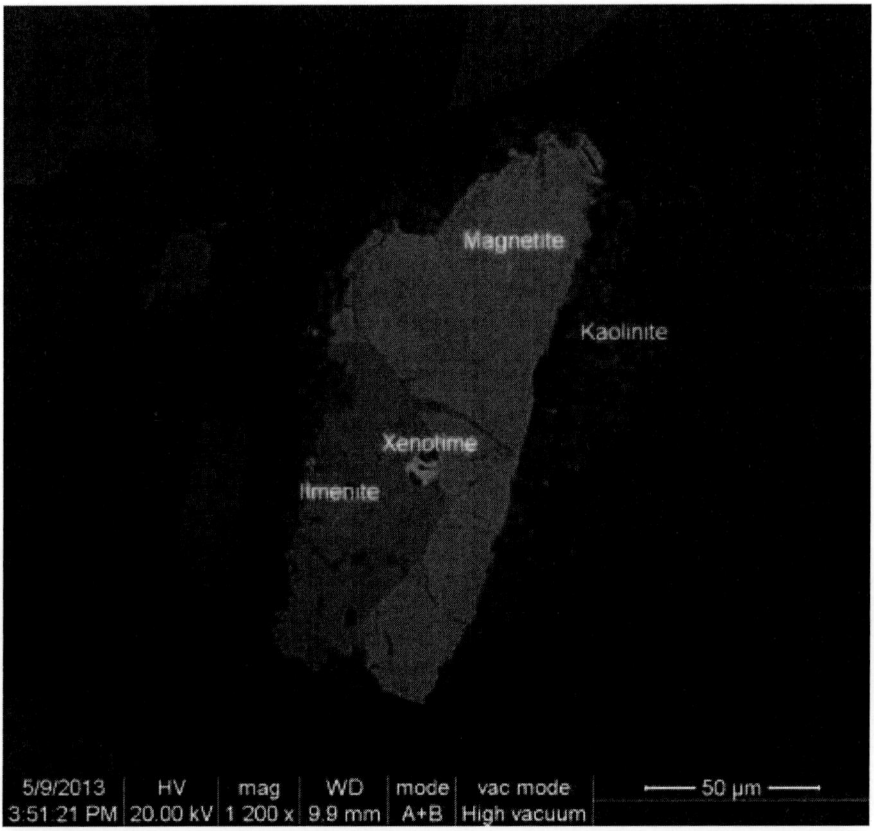

Figure 5. SEM images of ilmenite.

3.2. Pre-Treatment of the Tailings

The data in Table 2 show that most of the mineral in the tailing can be utilized in ceramic industry except of Fe_2O_3. Therefore, the pre-treatment of the tailings before re-utilization is mostly iron removal.

3.2.1. Fe Element Distribution

Kaolinite and quartz together make up 91.98% of the ion-adsorption rare earth tailings. It implies that the tailings can be re-utilized in ceramic application if the Fe content is less than 1.8% according to the requirement of TC-3 grade in the national standard GB/T 14563-1993. However, the content of Fe_2O_3 in the tailings is 2.11%, thus, further treatment to remove the iron is required.

Table 3 shows the distribution of Fe element in the tailings. It shows that titanomagnetite, kaolinite and limonite are the main Fe-bearing minerals in the tailings. Only titanomagnetite and limonite can be removed by magnetic separator methods. If titanomagnetite and limonite are removed, the Fe content in the tailings decreased to 1.23%, which meets the requirement of the standard of TC-3 grade for ceramic industry applications.

To ascertain the various particle sizes, the particle size distribution of the tailings is characterized by standard sieves method. Table 4 shows the particle size distribution and the grade of SiO_2, Al_2O_3, Fe_2O_3 and TiO_2 in the distribution. The +0.4 mm and −0.043 mm fraction are the main of the particle size distribution. The grade of SiO_2 decreases with decreasing particle size. The grade of SiO_2 in the +0.4 mm fraction is larger than 91.46%, implying that coarse grains are quartz and they can be separated by 0.4 mm sieve. Correspondingly, the fine grains (<0.043 mm) are kaolinite because the grade of Al_2O_3 increases with decreasing particle size. The change of the Fe_2O_3 grade is in accord with that of Al_2O_3. Considering the discussion in section 3.1, it can be concluded that the iron-bearing minerals is mostly mixed with kaolinite. The methods of iron removal from kaolinite reported in reference [8] -[11] can be applied in the tailings.

Table 3. Distribution of Fe element in the tailings.

Mineral	Mineral content in tailings /wt%	Fe content in minerals/ wt%	Fe content in tailings/ wt%	Ratio of Fe distribution/ wt%
Kaolinite	42.29	0.57	0.24	16.99
Quartz	49.69	0.17	0.08	5.98
Feldspar	3.90	0.16	0.01	0.43
Titanomagnetite	1.10	66.28	0.73	51.73
Limonite	0.29	53.19	0.15	10.85
Ilmenite	0.13	32.07	0.04	2.98
Muscovite	0.68	2.24	0.02	1.08
Biotite	1.36	10.34	0.14	9.98
Total	99.45	-	1.41	100

Table 4. Particle size distribution of the tailings and the quality of SiO_2, Al_2O_3, Fe_2O_3 and TiO_2.

Particles size /mm	Yield /%	Grade/%				Recovery rate/%			
		SiO_2	Al_2O_3	Fe_2O_3	TiO_2	SiO_2	Al_2O_3	Fe_2O_3	TiO_2
0.8	19.1	98.44	0.91	0.31	0.05	23.99	0.99	2.96	2.85
−0.8 ~ +0.4	7.45	91.46	6.24	1.22	0.15	8.69	2.65	4.54	3.34
−0.4 ~ +0.25	3.49	79.55	13.28	4.02	0.59	3.54	2.64	7.01	6.15
−0.25 ~ +0.12	7.01	74.19	16.99	3.81	0.62	6.64	6.79	13.35	12.98
−0.12 ~ +0.075	4.15	72.70	18.90	3.30	0.72	3.85	4.47	6.85	8.93
−0.075 ~ +0.043	4.81	73.17	19.41	2.91	0.76	4.49	5.33	7.00	10.92
−0.043	53.99	70.84	25.04	2.16	0.34	48.80	77.12	58.29	54.83
∑(Crude tailing)	100	78.32	17.50	2.11	0.38	100	100	100	100

3.2.2. Iron Removal by Magnetic Separator

The −0.4 mm fraction of the tailings is treated by a periodic high-gradient magnetic separator (SSS-I-145). The Fe content of the sorted tailings is measure by XRF. The effect of the magnetic induction on the Fe content is listed in Table 5.

As can be seen, the Fe grade of the sorted tailings decreases with the increase of the magnetic induction. However, the ratio of iron removal is not proportional to the magnetic induction. The ratio of iron removal increases quickly with the increase of the magnetic induction before the magnetic induction is 0.6 Tesla. And then, the ratio of iron removal increases slightly (4.68%) when the magnetic induction increases further from 0.6 T to 1 T. When the magnetic induction is 0.6 T, the Fe grade of the sorted tailings is 1.06 wt% and the ratio of iron removal is 50.41%, indicating most of titanomagnetite and limonite is removed. Therefore, the magnetic induction of 0.6 Tesla is more economical.

To remove the iron in kaolinite, redox method [9] [10] is attempted in the experiment. The −10 μm, −7 μm and −5 μm fractions of the tailings are obtained by settling method according to Stokes equation. The iron removal of the fine tailings is listed in Table 6.

The results in Table 6 show that the iron removal of the tailings by H_2O_2 treatment is insufficient in all the three fractions of the tailings, implying that most of Fe element is in the oxidized state. By using the $Na_2S_2O_4$ treatment, the Fe content of the −7 μm fraction changed from 1.68% to 1.27%, which is equivalent to 25.4% removal, indicating that $Na_2S_2O_4$ method is the good way to the iron removal of the fine tailings.

Table 5. Effect of the magnetic induction on the Fe content.

Magnetic induction/T	Fe_2O_3	
	Grade/wt%	Ratio of iron removal/%
0.2	1.40	34.78
0.4	1.26	41.02
0.6	1.06	50.41
0.8	0.99	53.80
1	0.96	55.09
Crude ore	2.11	-

Table 6. Iron removal of the fine tailings by redox method.

Tailings	Fe_2O_3 Content /wt%		
	Crude tailings	After H_2O_2 treatment	After $Na_2S_2O_4$ treatment
−10 μm	1.68	1.62	1.53
−7 μm	1.68	1.67	1.27
−5 μm	1.65	1.65	1.61

4. CONCLUSION

The ion-adsorption rare earth mining tailings in Southern China can be re-utilized because kaolinite and quartz share 91.98 wt% of the tailing. The other minerals include feldspar, muscovite, biotite, titanomagnetite, limonite and ilmenite, which are the important raw materials in ceramic industry. The particle size distribution of the tailings is bimodal. 91.46 wt% of the +0.4 mm fraction is quartz and the −0.043 mm fraction is mostly kaolinite. The two minerals can be separated by sieves. The total Fe content of the tailings is 2.11 wt%, which distributes mostly in titanomagnetite, kaolinite and limonite. The iron-bearing minerals can be removed by the periodic high-gradient magnetic separator with the magnetic induction of 0.6 T. The Fe content of the sorted tailings is less than 1.06 wt%, which meets the Chinese national standard of TC-3 grade for ceramic industry application. The Fe content in kaolinite can decrease further after $Na_2S_2O_4$treatment.

ACKNOWLEDGEMENTS

The authors gratefully acknowledge the financial support provided by the National High-tech Research and Development Program of China (Grant No. 2012AA061903).

REFERENCES

1. Su, W. (2009) Economic and Policy Analysis of China's Rare Earth Industry. China Financial and Economic Publishing House, Beijing. (In Chinese)
2. Chen, Z.H. (2011) Global Rare Earth Resources and Scenarios of Future Rare Earth Industry. Journal of Rare Earths, 29, 1-6.
3. Wübbeke, J. (2013) Rare Earth Elements in China: Policies and Narratives of Reinventing an Industry. Resources Policy, 38, 384-394.
4. Yang, X.J., Lin, A.J., Li, X.L., Wu, Y.D., Zhou, W.B. and Chen, Z.H. (2013) China's Ion-Adsorption Rare Earth Resources, Mining Consequences and Preservation. Environmental Development, 8, 131-136.
5. Guo, W. (2012) The Rare Earth Development Can No Longer Overdraw Ecological Cost. China Environment News, China Environmental Press, Beijing.
6. Sylvester, P. (2012) Use of the Mineral Liberation Analyzer (MLA) for Mineralogical Studies of Sediments and Sedimentary Rocks. Mineralogical Association of Canada Short Course Series, 42, 1-16. [Paper reference:1]
7. Chen, L. (2011) Effect of Magnetic Field Orientation on High Gradient Magnetic Separation Performance. Minerals Engineering, 24, 88-90.

8. 8. Svoboda, J. and Fujita, T. (2003) Recent Developments in Magnetic Methods of material Separation. Minerals Engineering, 16, 785-792.

9. Zegeye, A., Yahaya, S., Fialips, C.I., White, M.L., Gray, N.D. and Manning, D.A.C. (2013) Refinement of Industrial Kaolin by Microbial Removal of Iron-Bearing Impurities. Applied Clay Science, 86, 47-53.

10. Pham, A.L.T., Doyle, F.M. and Sedlak, D.L. (2012) Kinetics and Efficiency of H2O2 Activation by Iron-Containing Minerals and Aquifer Materials. Water Research, 46, 6454-6462.

11. Santos, E., Scorzelli, R.B., Bertolino, L.C., Alves, O.C. and Munayco, P. (2012) Characterization of Kaolin from the Capim River Region—Brazil. Applied Clay Science, 55, 164-167.

12. Chandrasekhar, S. and Ramaswamy, S. (2007) Investigation on a Gray Kaolin from South East India. Applied Clay Science, 37, 32-46.

13. Xia, G., Lu, M., Su, X. and Zhao, X. (2012) Iron Removal from Kaolin Using Thiourea Assisted by Ultrasonic Wave. Ultrasonics Sonochemistry, 19, 38-42.

7

CHAPTER

Normalized Rare Earth Elements in Water, Sediments, and Wine: Identifying Sources and Environmental Redox Conditions

David Z. Piper[1], Michael Bau[2]

[1]*Geological Survey, Menlo Park, USA*

[2]*Jacobs University, Bremen, Germany*

ABSTRACT

The concentrations of the rare earth elements (REE) in surface waters and sediments, when normalized on an elementby-element basis to one of several rock standards and plotted versus atomic number, yield curves that reveal their partitioning between different sediment fractions and the sources of those fractions, for example, between terrestrial-derived lithogenous debris and seawater-derived biogenous detritus and hydrogenous metal oxides. The REE of ancient sediments support their partitioning into these same fractions and further contribute to the identification of the redox geochemistry of the sea water in which the sediments accumulated. The normalized curves of the REE that have been examined in several South American wine varietals can be interpreted to reflect the lithology of the bedrock on which the vines may have been grown, suggesting limited fractionation during soil development.

Keywords: Rare Earth Elements; Cerium Redox; Seawater/ Sediment; Sources

1. INTRODUCTION

The rare earth elements (REE), or lanthanide elements, are a group of 14 elements that have similar crystal ionic radii and valance state. They have a decreasing crystal ionic radius from La (1.016 Å) to Lu (0.85 Å) and a single 3+ valence within the siliciclastic fraction of sediments and sedimentary rocks. This imposes limitations on their fractionation within siliciclastic debris during weathering and eventual transport by rivers as a suspended or colloidal fraction. Thus, they are an ideal group to identify the sources of the lithogenous fraction of marine and non-marine sedimentary deposits, e.g., maficand felsic-igneous and sedimentary rocks.

By contrast, the REE in the dissolved load of rivers may exhibit measurable fractionation [1-3]. Cerium, in the 3+ valence state in the suspended load of rivers, is oxidized to the highly insoluble 4+ valence state that continues upon its entering the ocean [4-9]. As a result, Ce is depleted in seawater, relative to the remaining elements of the group that maintain their 3+ valence state. Nonetheless, the 3+ valence REE also exhibit interelement fractionation, owing to their formation of complexes mostly with CO_3^{2-} [10-12]. In seawater, the complexed species exhibit an increase in their stability with increasing atomic number. This results in a relative increase in their concentration with increasing atomic number. Adsorption reactions on particle surfaces contribute further to their fractionation [13].

These differences in the behavior of the individual REE in seawater are recorded by differences in the REE curves of the various seawater-derived fractions of marine sediments, i.e., the marine hydrogenous and biogenous sediment fractions [4] versus the terrestrial-derived lithogenous fraction. For example, abyssal ferromanganese nodules (e.g., [6,14]) and seamount ferromanganese crusts [15-18] exhibit an enrichment of Ce, whereas pelletal carbonate fluorapatite (CFA), glauconite [19], phillipsite [20], biogenic calcium carbonate [21], opal [6], and less so barite [22] exhibit a Ce depletion relative to its nearest neighbors La and Pr. The 3+ valence REE commonly exhibit an increase in concentration with increasing atomic number, again a reflection of their propensity to form complexes with CO_3^{2-} and other ligands. We review here a fraction of the research on the behavior of the REE in soils, surface waters, sediments, and sedimentary rocks; from source rocks, to streams, to seawater, to their deposition and eventual burial in sea floor sediments. They have proven to be invaluable in contributing to an overall improved understanding of the geochemistry of earth-surface processes.

2. ANALYTICAL TECHNIQUES

The REE gained wide attention following the work of Haskin and Gehl [23], Goldberg et al. [4], and Høgdahl et al. [24], who documented the potential that these elements offered in advancing our understanding of the geochemical history of sedimentary environments. They showed that several members of the group can be measured by neutron activation analysis in marine sediments and seawater, in which the concentrations are as low as a few nanograms per liter (in seawater) to micrograms per gram (in sediments). This method was expanded to include the examination of a wide range of sedimentary components following the development of high-resolution Ge(Li) g-ray detectors, e.g., soils, marine and nonmarine sediments, volcanic deposits, biogenic detritus, aerosols, the dissolved and suspended fractions of steams and seawater, moon rocks, and chondrites. Instrumental neutron activation analysis remained widely used for some 20 years until it was largely supplanted by other techniques, coupled with various leaching and extraction procedures that have allowed for the analysis of the complete group of REE in a wide range of sample types [1,5,8,9,16,25]. These techniques also greatly improved the analytical accuracy and precision, now better than ca 5%.

The REE are the prime example of the Oddo-Harkins Rule [26-28], whereby even-numbered elements are more abundant than odd-numbered elements (**Figure 1**), owing to the greater nuclear stability gained by the pairing of protons that offsets the spin of the one by the other. Promethium (atomic number 61) does not occur

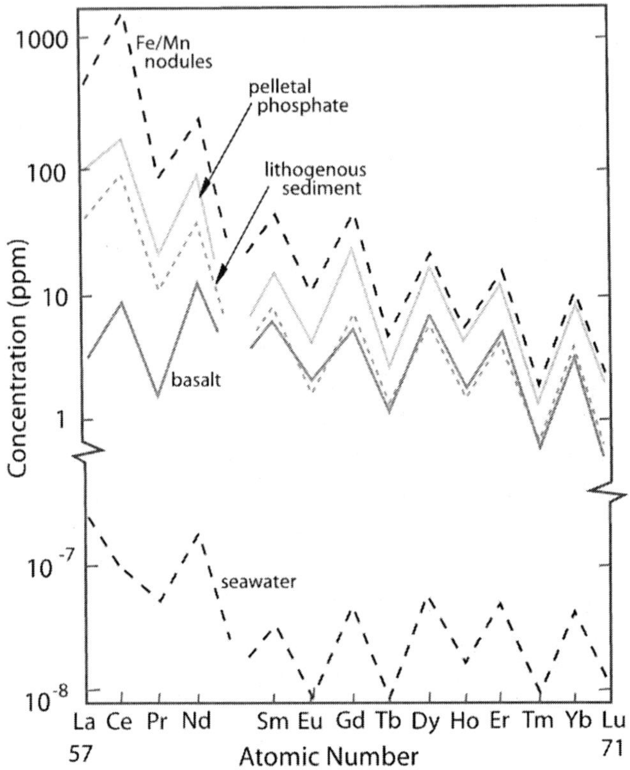

Figure 1. The measured concentrations of REE in seawater, several marine sediment phases, and basalt, all of which exhibit the odd/even concentration difference explained by the Oddo-Harkins Rule.

naturally, having no isotope with a half-life greater than ca 20 days. The herringbone-shaped curve of their measured concentrations, when plotted versus atomic number, is converted to a smooth curve by normalizing the REE on an element-by-element basis to one of several rock standards (**Figure 2**). The procedure allows the relative concentrations of the REE in different samples to be easily compared visually.

Standards that are commonly used (**Table 1**) include the World Shale Average (WSA), as calculated by Piper [20] from analyses published by Haskin and Haskin [29] and Wedepohl [30]; the North American Shale Composite (NASC), analyzed by Gromet et al. [31];

the Upper Continental Crust (UCC), with several slightly different values reported by numerous individuals (e.g., [32-34]), but with quite similar interelement concentrations; Post Archean Australian Shale (PAAS), advanced by McLennan [35]; and lastly an average of chondrites [26]. The concentrations of the REE in these standards represent two compositional extremes of siliciclastic-source-rocks (**Figure 2**), the one felsic (WSA, UCC, PAAS, NASC) and the second ultramafic (chondrites).

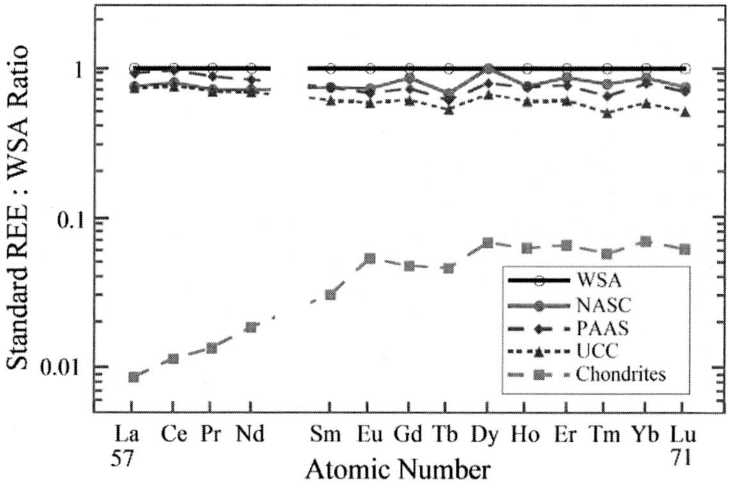

Figure 2. The REE curves of rock standards (**Table 1**): World Shale Average (WSA), Post Archean Australian Shale (PAAS), North American Shale Composite (NASC), the Upper Continental Crust (UCC), and an average of chondrites.

Although the curves for the WSA and NASC exhibit slightly higher concentrations of the heavy REE versus the light REE, than the curves for the UCC and PAAS, these standards exhibit normalized curves that are quite similar. Considering that they represent sedimentary rocks of different geologic ages that were collected throughout the world and likely have been recycled numerous times (from surface weathering under various conditions, to erosion, to stream and/or wind transport, to ocean deposition under possibly different geochemical environments, to burial, to uplift, to...), their virtually identical interelement REE concentrations seem to reflect a

Table 1. Elemental composition of rock standards—World Shale Average (WSA), North American Shale Composite (NASC), Upper Continental Crust (UCC), Post Archean Australian Shale (PAAS), and average chondrites.

Rare Earth Element	World Shale Average [20]	N. American Shale Composite [31]	Upper Continental Crust [32-34]	Post Archaen Australian Shale [35]	Average Chondrites [26]
La	41.00	31.1	30	38.2	0.32
Ce	83.00	66.7	64	79.6	0.90
Pr	10.10	7.70	7.1	8.83	0.13
Nd	38.00	27.4	26	33.9	0.57
Sm	7.50	5.59	4.5	5.55	0.21
Eu	1.61	1.18	0.88	1.08	0.074
Gd	6.35	4.90	3.8	4.66	0.31
Tb	1.23	0.85	0.64	0.774	0.051
Dy	5.50	4.17	3.5	4.68	0.30
Ho	1.34	1.02	0.8	0.991	0.074
Er	3.75	2.84	2.3	2.85	0.21
Tm	0.63	0.48	0.33	0.405	0.032
Yb	3.53	3.06	2.2	2.82	0.18
Lu	0.61	0.46	0.32	0.433	0.032

coherence for this group of chemical elements quite beyond what might have been expected, considering their different possible valence states, slightly different ionic crystal radii in the 3+ valence state, and varying propensity to form complexes in solution with various ligands.

Goldberg et al. [4], Haskin and Haskin [29], and Høgdahl [36] employed the average of several chondrites to normalize the REE in marine samples they examined, although it is perhaps more appropriately used in the study of mafic and ultramafic rocks. Piper [20] proposed that the WSA represents perhaps a more appropriate standard in the examination of marine and non-marine sediments and surface waters, owing to similar relative concentrations of the REE in the standard and their concentrations in sediments, river waters, and less so seawater. The WSA continues to be used, but the UCC and PAAS standards are perhaps now more commonly used. The chondrite standard is still used by some in their analysis of marine and non-marine sediments and surface water, although possibly more to mask analytical uncertainties than to reveal the similar behavior of the REE. Goldstein and Jacobsen [1] obtained rather different values for the NASC than Gromet et al. [31], which they attributed largely to heterogeneity of the standard. An average of a subgroup of samples being examined has also been employed to normalize individual sets of samples as it can emphasize subtle differences between samples that might otherwise escape detection. The standard one elects to use, however, is quite unimportant, as long as its interelement concentrations are similar to those in the samples being examined.

3. REE IN RIVERS AND SEAWATER

3.1. Rivers

The REE in several of the world's major rivers [37], normalized to any one of the felsic rock standards, reveal that the bulk REE concentrations exhibit a rather flat curve (**Figure 3**(a)). In contrast to these curves, Goldstein and Jacobsen [1], Elderfield et al. [2], and Sholkovitz [3] established that the dissolved load of rivers, although representing a minor fraction of the total REE load of rivers, is fractionated from the suspended load, which is dependent on the chemistry of the water (e.g., salinity, pH, alkalinity, and biological activity). In general, the result of the fractionation is to increase the heavy REE concentrations in the dissolved fraction, relative to the light REE concentrations (**Figure 3**(b)). This fractionation between the dissolved and suspended load was interpreted to contribute directly to the REE curve of the dissolved fraction of seawater, but reactions between seawater and river water within estuaries greatly complicate the contribution of river borne REE entering the ocean to the eventual composition of seawater.

The flat curve of the suspended load of rivers commonly resembles the curve of the source rocks [2] in the drainage basin (**Figure 4**(a)). It contrasts sharply with the curves and elevated concentrations of the bed load (Figure 4(b)) of streams [38]. The concentrations of the heavy-mineral-hosted elements—Ti and Zr—also exhibit strong enrichments within the stream bed load with their order of enrichment and that of the REE being the following:

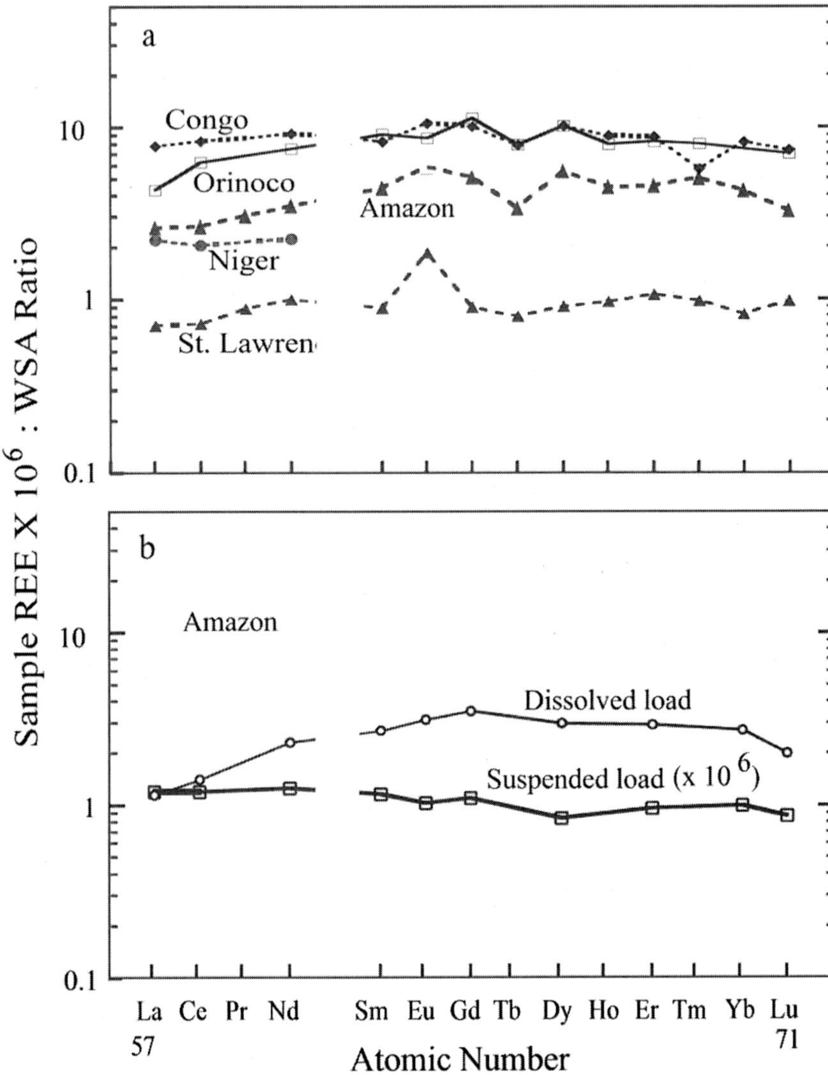

Figure 3. (a) The REE curves of several of the world's major rivers [37]; (b) The curves of the dissolved and suspended fractions in the Amazon River, separated by filtration [3].

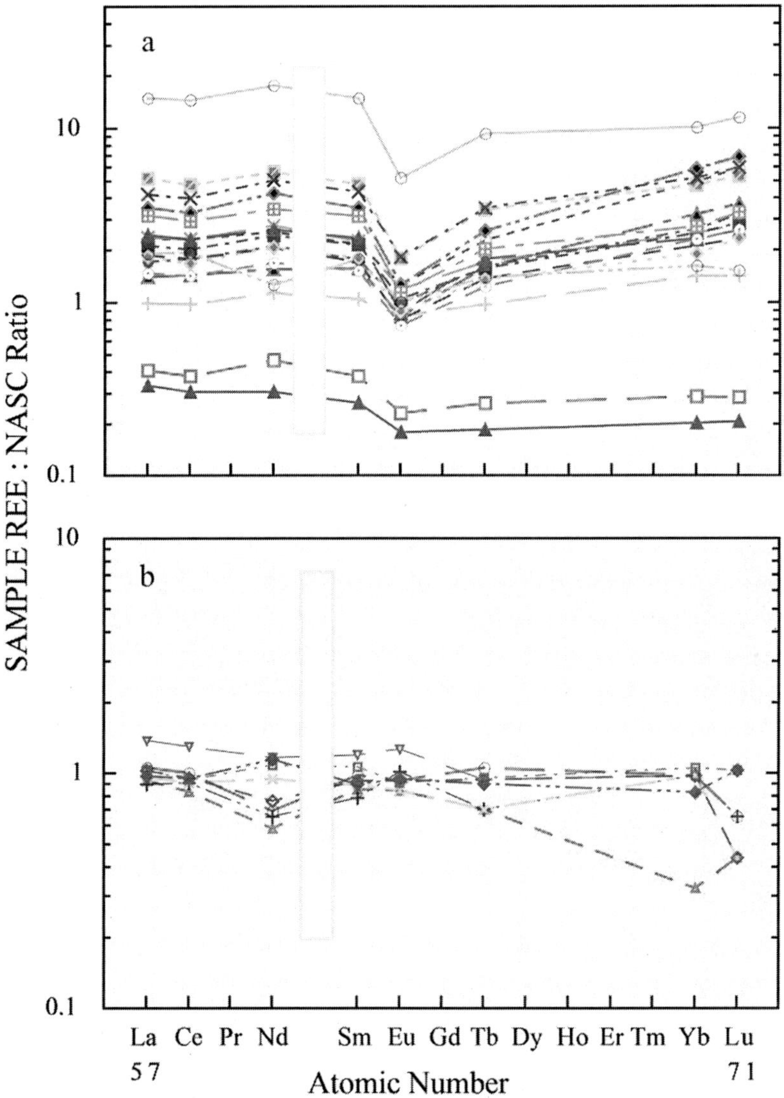

Figure 4. (a) The REE curves of bed sediments in the Big Black River [38], a tributary to the Mississippi River and (b) of soils within its drainage basin.

$$\text{bed sediment} \square \ \text{soil} \geq \text{suspended sediment} \approx \text{source rock} \approx \text{WSA} \quad , \tag{1}$$

suggesting that the heavy minerals are a major host of the REE in bed sediments.

3.2. Seawater

In contrast to the partitioning of the REE largely into the suspended fraction in rivers [1], they are strongly partitioned into the dissolved fraction in seawater, i.e., into filtered seawater. The curve of seawater, in general, exhibits a negative Ce anomaly [4,5,8,9,24,39,40] and progressive enrichment of the heavy REE, relative to the light REE (Figures 5 and 6(a)). Although the curves are somewhat different between the oceans and with depth in each individual ocean, the changes in the normalized curves with depth clearly reflect the exchange reactions between suspended phases and the dissolved load that are superimposed on the history of advection of the different water masses. The water masses in the Atlantic Ocean, with increasing depth, are surface water, Antarctic Intermediate Water, North Atlantic Deep Water, and Antarctic Bottom Water [41]. The advection of Antarctic Bottom Water that has relatively low REE concentrations produces a mid-water REE concentration maximum in both the Atlantic and Pacific Ocean [8,36].

The negative Ce anomaly of seawater reflects its oxidation to the strongly insoluble 4+ valence under oxic to suboxic redox conditions in the open ocean today (**Table 2**), in contrast to the other REE that maintain a 3+ valence. They too are fractionated from each other by their complexation with CO_3^{2-} and less so with HPO_4^{2-}, and adsorption on the surfaces of suspended particles [12,13]. The complexes exhibit an increase in stability with in creasing REE atomic number, i.e., the heavy REE are more strongly partitioned into the complexed fraction versus the ionic state than are the light REE. As the complexed fraction of each REE is less reactive to adsorption reactions, the heavy REE are less readily removed from seawater onto suspended phases than are the light REE [3]. Thus, the REE exhibit an increase of their normalized concentrations in the seawater with increasing atomic number, as reflected by their increase in residence time in the ocean—500 yrs for La increasing to 2890 yrs for Lu [57]. Cerium, with a residence time of only 50 yrs, is an exception to this trend, owing to its oxidation to the 4+ valence (**Table 2**).

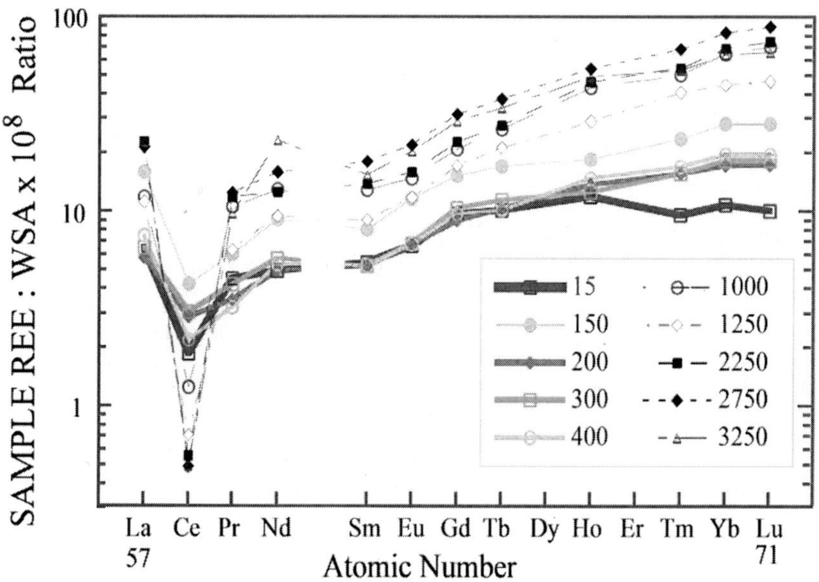

Figure 5. The REE curve of seawater from the eastern Pacific Ocean [5]. Note the REE curve of the samples from 200 to 300 m water depth, from the core of the O_2 minimum layer, with a normalized Ce concentration that is less than the Ce anomaly of waters collected at both shallower and deeper depths.

Separation of the suspended load of seawater, the fraction > 0.4 µ that is recovered by filtration [9], from the dissolved load reveals the partitioning of the REE within the suspended fraction largely into an Al-silicate lithogenous fraction, a Mn-oxide fraction, and likely a Fe-oxide fraction. The composition of each of the suspended fractions contrasts with the dissolved fraction (**Figure 6**). The Mn-oxide fraction exhibits a strongly evolving curve with ocean depth. Within

Table 2. Half-cell reactions assuming standard-state conditions and seawater concentrations for the trace elements. The master half-cell reactions for seawater are the three reactions written in bold print, owing to the elevated concentration O2, NO_3^-, and SO_4^{2-} above those of the trace metals. Ground waters may have these same master half-cell reactions, but concentrations will likely be quite different.

Eh	Half-cell reactions	A#	Seawater concentration in moles/kg, or partial pressure	Ref.*
1.099	$3Co^{2+}(aq) + 4H_2O \rightarrow Co_3O_4(s) + 2e^- + 8H^+(aq)$	6	$Co^{2+} = 2.04 \times 10^{-11}$	[42,43]
0.805	$\mathbf{2H_2O(l) \rightarrow O_2(g) + 4e^- + 4H^+(aq)}$	- -	$\mathbf{O_2 = 0.2\ atm}$	**[44]**
0.704	$Mn^{2+}(aq) + 2H_2O(l) \rightarrow MnO_2(s) + 2e^- + 4H^+(aq)$	6	$Mn^{2+} = 2.5 \times 10^{-10}$	[44,45]
0.702	$I^-(aq) + 3H_2O(l) \rightarrow IO_3^-(aq) + 6e^- + 6H^+(aq)$	3	$I^- = 5.0 \times 10^{-9}$ $IO_3^- = 4.5 \times 10^{-7}$	[42,44]
0.698	$\mathbf{N_2(g) + 6H_2O(l) \rightarrow 2NO_3^-(aq) + 10e^- + 12H^+(aq)}$	**3**	$\mathbf{N_2 = 0.8\ atm}$ $\mathbf{NO_3^- = 3.9 \times 10^{-5}}$	**[44,46]**
0.545	$Cr(OH)_3(s) + H_2O(l) \rightarrow CrO_4^{2-}(aq) + 3e^- + 5H^+(aq)$	4	$CrO_4^{2-} = 4.04 \times 10^{-9}$	[43,47,48]
0.448	$SeO_3^{2-}(aq) + 2H_2O(l) \rightarrow SeO_4^{2-}(aq) + 2e^- + 2H^+(aq)$	4	$\sum Se = 2.22 \times 10^{-9}$ $SeO_3^{2-} = 0.92 \times 10^{-9}$	[44,49,50]
0.320	$Ce^{3+}(aq) + 2H_2O(l) \rightarrow CeO_2(s) + e^- + 4H^+(aq)$	9	$Ce^{3+} = 1.0 \times 10^{-11}$	5, 44
0.296	$Fe^{2+}(aq) + 3H_2O(l) \rightarrow Fe(OH)_3(s) + e^- + 3H^+(aq)$	6	$Fe^{2+} = 1.20 \times 10^{-9}$	44, 45
0.121	$Se(s) + 3H_2O(l) \rightarrow SeO_3^{2-}(aq) + 4e^- + 6H^+(aq)$	4	$SeO_3^{2-} = 1.11 \times 10^{-9}$	49
0.013	$UO_2(s) + 2HCO_3^-(aq) \rightarrow UO_2(CO_3)_2^{2-}(aq) + 2e^- + 2H^+(aq)$	4, 5	$HCO_3^- = 2.47 \times 10^{-3}$ $UO_2(CO_3)_2^{2-} = 1.26 \times 10^{-8}$	[44,45,51]
−0.018	$2Fe^{2+}(aq) + 3H_2O(l) \rightarrow Fe_2O_3(s) + 2e^- + 6H^+(aq)$	6	See above	[44,45]
−0.040	$V_2O_4(s) + 4H_2O(l) \rightarrow 2H_2V_2O_4^-(aq) + +2e^- + 4H^+(aq)$	3	$H_2V_2O_4^- = 4.0 \times 10^{-8}$	[43,52]
−0.055	$\mathbf{HS^-(aq) + 4H_2O(l) \rightarrow SO_4^{2-}(aq) + 8e^- + 9H^+(aq)}$	**3, 4**	$\mathbf{HS^- = 7.75 \times 10^{-28}}$ $\mathbf{SO_4^{2-} = 2.8 \times 10^{-2}}$	**[44,46]**
−0.055	$Cu_2S(s) + 4H_2O(l) \rightarrow SO_4^{2-}(aq) + 2Cu^+(aq) + 8e^- + 8H^+(aq)$	4, 6	$Cu^+ = 3.59 \times 10^{-9}$	[42,45,53]
−0.137	$CdS(s) + 4H_2O(l) \rightarrow SO_4^{2-}(aq) + Cd^{2+}(aq) + 8e^- + 8H^+(aq)$	4, 6	$Cd^+ = 6.94 \times 10^{-10}$	[42,44,46,54]
−0.162	$ZnS(s) + 4H_2O(l) \rightarrow SO_4^{2-}(aq) + Zn^{2+}(aq) + 8e^- + 8H^+(aq)$	4, 6	$Zn^{2+} = 5.97 \times 10^{-9}$	[42,44]
−0.175	$MoS(s) + 12H_2O(l) \rightarrow MoO_4^{2-}(aq) + 2SO_4^{2-}(aq) + 18e^- + 24H^+(aq)$	4, 5	$MoO_4^{2-} = 1.1 \times 10^{-7}$	[44,55,56]
−0.188	$NiS(s) + 4H_2O(l) \rightarrow SO_4^{2-}(aq) + Ni^{2+}(aq) + 8e^- + 8H^+(aq)$	4, 6	$Ni^{2+} = 8.01 \times 10^{-9}$	[42,44,46,53]
−0.209	$FeS(s) + 4H_2O(l) \rightarrow SO_4^{2-}(aq) + Fe^+(aq) + 8e^- + 8H^+(aq)$	4, 6	$Fe^{2+} = 1.20 \times 10^{-9}$	[42,44,45,53]
−0.210	$UO_2(s) + 3HCO_3^-(aq) \rightarrow UO_2(CO_3)_3^{4-}(aq) + 2e^- + 3H^+(aq)$	4, 5	$UO_2(CO_3)_3^{4-} = 1.26 \times 10^{-8}$	[42,44,45]

#Values of the effective ionic diameter used in the Debye-Hückel equation.

the upper ca 100 m of the water column, it exhibits a negative Ce anomaly (**Figure 6**(b)), similar to the dissolved fraction of seawater itself (Figure 6(a)). The anomaly becomes strongly positive to ca 1000 m depth, possibly in response to the crystallization of the fully oxidized suspended Mn oxide, on which Ce^{4+} is adsorbed through biologically mediated redox and adsorption reactions [58]. The 3+

valence REE are also enriched in the Mn-oxide, the light REE more so than the heavy REE in accord with their partition coefficients with CO_3^{2-} complexes [11-13]. Their partitioning into both Fe and Mn fractions of ferromanganese crusts that are universally present on ocean seamounts suggests that both Mn and Fe oxides contribute to the depth curves of the REE in the suspended load [16,18]. Thus, the curves for the REE in seawater are established by their partitioning between dissolved and suspended fractions within seawater itself, that is, by their complexation mostly with CO_3^{2-} that controls adsorption/desorption reactions onto and off of particle surfaces, redox reactions mitigated by the oxidation of organic matter, and the presence of a geochemically largely non-reactive Al-silicate fraction.

4. REE IN SEDIMENTS

4.1. Terrestrial Sediment

Loess deposits may perhaps represent a reliable average composition of terrigenous sedimentary deposits. Taylor et al. [59] reported that these deposits, collected from several continents, have REE curves (**Figure 7**) that closely resemble the shale curve (**Figure 2**), with slight fractionation between light and heavy REE as a function of grain size [60] and surface weathering [61]. Their bulk composition might, however, be represented by a single grain size, by the <4 μ fraction [62]. The REE curves of the eolian deposits, thus, are strongly reflective of the mean curve of their source rocks.

4.2. Marine Sediment Faction

Marine sediments that accumulated distant from major volcanic activity are composed of essentially two fractions, a land-derived lithogenous fraction and a seawaterderived fraction, which is sub-divided into a hydrogenous fraction and a biogenous fraction [4]. Aluminum serves as a proxy for the lithogenous fraction [63,64]. Examples of trace elements also hosted predominantly by this fraction include the alkali metals, heavy-mineral hosted trace elements, plus Nb, Ga, Sc, Ta as evidenced by their linear correlations with Al that extrapolate essentially to the origin [65]. Their relationship with Al adds support to the use of Al alone as a proxy for this sediment fraction. The lithogenous contribution of the REE can then be estimated from the REE:Al ratios of one of the rock standards whose Al-normalized metal contents, e.g., the alkali-metal:Al ratios, are closest to those of the sediments being examined.

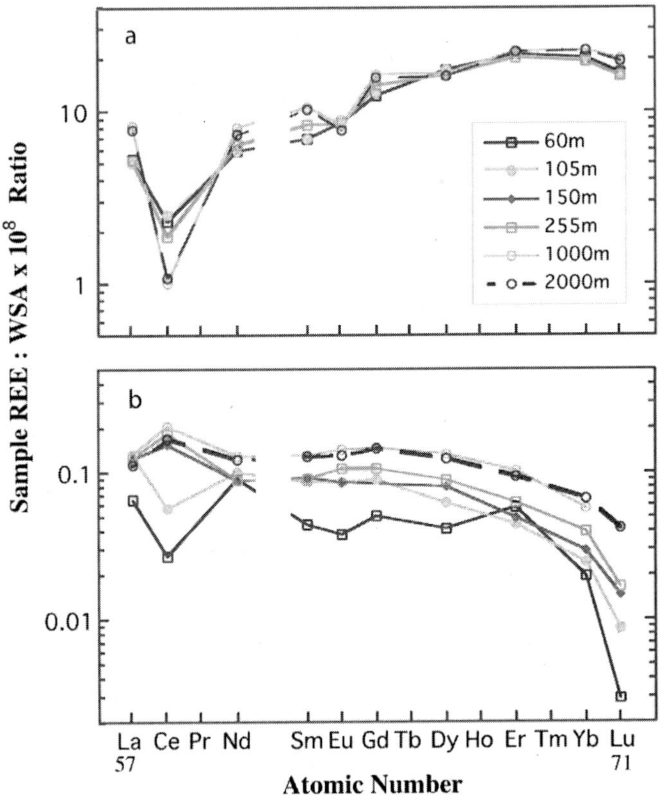

Figure 6. The REE curve of seawater from the Sargasso Sea in the Atlantic Ocean [7] and (b) of the suspended fraction of seawater collected on a 0.4 μm filtered and dissolved in acetic acid [9].

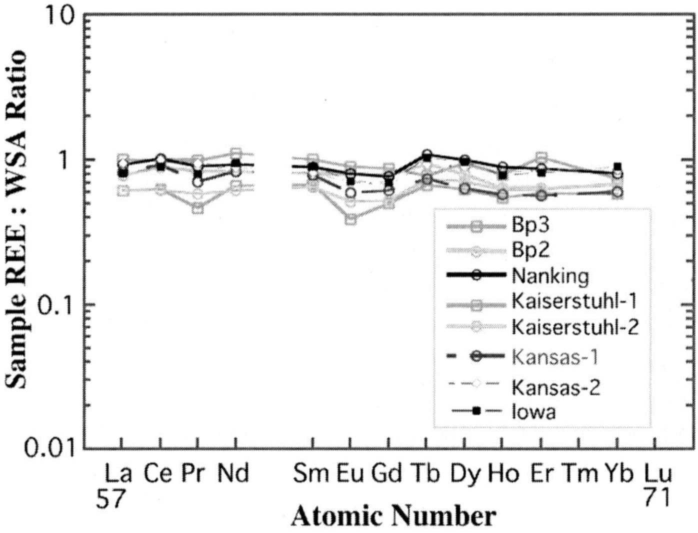

Figure 7. The REE curves of loess deposits. See Taylor et al. [59] for sample locations.

The marine fraction of the REE is then estimated by the elevated REE:Al ratios above the ratios in the rock standard. This fraction may further include C as organic matter, Ca and Si as $CaCO_3$ and opal tests of plankton, respectively; Cd, Cu, and Ni as trace nutrients; Cr, Fe, Mn, Mo, U, and V, plus Cd and Cu [66] as redox sensitive trace elements; and inorganically and biologically mediated surface-adsorbed elements. Along several ocean margins that exhibit relatively low accumulation rates of lithogenous sediment and suboxic conditions in the water column, i.e., NO_3^- reduction or denitrification (**Table 2**), a seawater-derived REE fraction can be present above the terrigenous input of sedimentary debris. Similarly, sediments from the oxic pelagic environment can also exhibit REE concentrations that are elevated above a lithogenous contribution. Sediments that accumulate under seawater anoxic sulfate-reducing conditions, for example, that typify the bottom waters of the Black Sea and Cariaco Basin, do not exhibit a seawater-derived REE fraction, as evidenced by the REE profiles in the water column of these basins [67-69] and the REE:Al ratios of the basin sediments [65].

4.2.1. Lithogenous Sediment Fraction

Marine sediments that have accumulated along ocean margins, where the lithogenous sediment fraction dominates the bulk sediment composition, commonly exhibit REE curves that represent the composition of their terrigenous source rocks [27,70-73]. Modeling and leaching analyses of Quaternary pelagic sediments from the central Pacific Ocean reveal that the siliciclastic fraction of these sediments similarly has a flat WSA-normalized curve [14].

The relationship between the REE composition of sediments and their terrigenous sources is further recorded by the sediments that are accumulating under the strongly negative redox condition of the Black Sea [65]. Two major terrigenous sources of significantly different composition were identified initially by the clay mineralogy of the surface sediments [74]. In the northern region of the basin, the fine fraction of the sediments is dominated by illite. Its distribution identifies its source as largely Eastern Europe, delivered to the basin mostly by the Danube River and rivers along the northern and western margin of the basin. In the southern region, the clay fraction is enriched in montmorillonite (smectite) that is derived largely from Anatolia. The major-element composition of the suspended phases in streams draining the two regions [74,75] further defines differences in the elemental composition of the two source regions.

The REE in surface sediments of the Black Sea [65] support the results of the mineralogy studies. Surface sediments recovered in box cores from the western basin margin have a flat REE curve that is representative of the largely felsic rocks of eastern Europe, whereas surface sediments from the southern basin margin have a light REE depletion that reflects a moderate mafic-rock contribution from the ultramafic rocks of the Pontic Mountains in northern Anatolia. Although the dominant sediment source currently is the river input from eastern Europe [74], the Anatolia source can be appreciated when it is realized that the Pontic Mountains rise sharply from a narrow 5 to 10 km wide coastal plain to an elevation of ca 3950 m. The coastal plain gives way seaward to an equally narrow basin shelf and slope, the latter traversed by numerous submarine canyons that occur along the full extent of the Anatolia slope [76]. This system of canyons likely delivers much of the sediment from the Pontic Mountains to the abyssal plain by by-passing the shelf and slope, possibly more so in the past than presently.

In support of this interpretation, the sediments that accumulated on the abyssal plain during the lacustrine phase of the Black Sea, following the Last Glacial Maximum (LGM) but prior to the invasion of Mediterranean water through the Bosporus 9.3 ka, i.e., when the Black Sea was a lake, exhibit the two different REE curves noted above. Within sediment cores recovered from the abyssal region of the basin, the curves change abruptly every few 10 s of centimeters (**Figure 8**), or ca 2000 to 3000 yrs [65], from the one type of curve to the other. The shift of the curves reflects a shift in the dominance of the northern and western source versus the southern and eastern source that was likely driven by regional changes of climate since the LGM [77].

4.2.2. Seawater Derived Fractions

Owing to the complexation of the 3+ REE in seawater [11,13] and the oxidation of Ce^{3+} to the 4+ valence [3,7,8, 78,79], the REE curves of different marine sediment fractions might be expected to reflect the alkalinity and redox potential of the seawater from which they accumulated. Although both features are strongly reflected in several individual components of modern deposits (Figure 9), e.g., phillipsite [14], biogenic carbonate [21] and opal [6], CFA [19,80],and possibly barite [22]; an exact one-to-one relationship with the seawater at the depth of accumulation is unclear. An exception is the pelletal CFA that is currently accumulating on the continental shelf of Peru. Here the apatite has a negative Ce anomaly and heavy REE enrichment that closely approach those of the seawater (Figures 5 and 9) at ca 150 to 300 m [5], the depth of accumulation of the CFA examined. The depth corresponds to the depth of the oxygen minimum zone (OMZ) of the eastern tropical Pacific [81], the lens of water at intermediate depth that is characterized by suboxic respiration, i.e., denitrification (**Table 2**).

Ferromanganese nodules on the abyssal sea floor (6, 14, and others) and seamount ferromanganese crusts [15- 18,82] are the major exception to this relationship between the REE composition of marine deposits and seawater (**Figure 9**). They have positive Ce anomalies, similar to the positive Ce anomaly of the suspended Mnoxide phase in the water column (**Figure 6**(b)) from greater than ca 1000 m depth [9], rather than the negative anomaly of the seawater dissolved load. The relationship of the 3+ valence REE in these deposits with their concentrations in seawater is equally

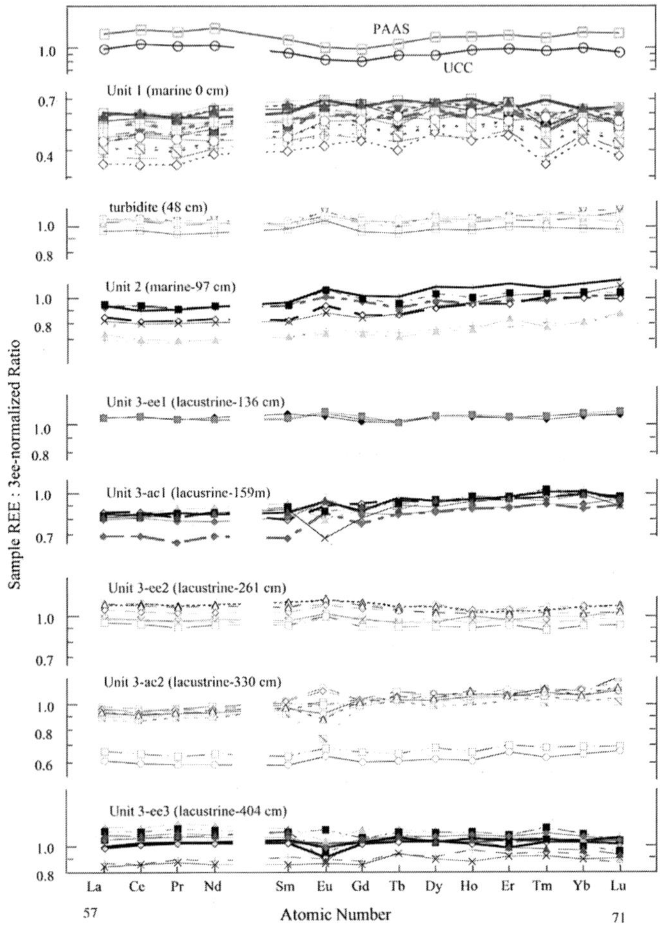

Figure 8. Rare earth elements in a sediment core from the central abyssal plain of the Black Sea [65], normalized to the average of the lacustrine sediments in the three "ee" subunits of Unit 3. The sediment core is divided into two marine units and the one lacustrine unit [76]. Subdivision of Unit 3 into 5 subunits is based on REE curves. The REE in the surface sediments of box cores from the western basin margin identify the "ee" subunits as having a dominantly eastern European source; the sediments of box cores from the southern margin identify the "ac" subunits as having an additional mafic source, the ultramafic rocks of the Pontic Mountains in northern Anatolia. The "ac" subunits exhibit a ca 15% to 20% depletion of the light REE, relative to the heavy REE. The upper boundary of each unit is given in the title; e.g., Unit 1 extends from 0 cm to 48 cm. The turbidite between Units 1 and 2 is present in many cores recovered throughout the basin. Its source was the Crimea slope [65]. The 3ee-normalized curves for the UCC and PAAS (**Table 1**) are included for reference.

complex, but it is likely determined largely by the partitioning of the REE into CO_3^{2-} complexes in the dissolved fraction of seawater [16].

5. CE OXIDATION IN ANCIENT OCEANS

Wright et al. [83] advanced the hypothesis that the REE curves of CFA in ancient phosphate deposits, specifically the negative Ce anomaly of biogenous apatite, might have recorded the redox chemistry of ancient oceans. They reasoned that the negativity of the Ce anomaly in seawater during periods of possible global anoxia (e.g., [84-87]) would be less than during times when the ocean was overwhelmingly oxic, as it is currently. Following their success, more recent work has failed to support the earlier findings [88]. Several reasons may account for this apparent failure, which may not necessarily reflect a failure of the original hypothesis. As noted earlier, the Ce anomaly of seawater becomes increasingly more negative from the surface to abyssal depths, with the major exception of its trend in the OMZ of the eastern Pacific Ocean (**Figure 5**). Further, the Ce anomaly within individual phosphate deposits commonly exhibits a range of values [80,89] that requires the analysis of several samples, particularly samples that are strongly enriched in CFA. Such samples may define the anomaly of CFA alone (**Figure 10**). And lastly, a diagenetic overprint may largely have altered the REE curve of ancient biogenic CFA following its deposition [90-93]. Thus, several factors complicate the composition of the apatite fraction of ancient deposits [94].

On the positive side, the pelletal CFA fraction of sedimentary deposits may largely retain its REE signature of deposition, similar to the retention of its C and O isotopic signatures [95,96]. Tens of samples of the pelletal CFAenriched units from the Phosphoria Formation of Permian age in southeast Idaho [80] and from the Lisburne Group of Carboniferous age in northern Alaska [97] exhibit Ce anomalies that are significantly more negative (Figures 10 and 11(a)) than the Ce anomalies of the Peru Shelf deposit (**Figure 9**). They average ca −0.9 in samples from these deposits with greater than 25% apatite. Analyses by us (DZP) of the pelletal CFA-deposit from the Tertiary Miocene Pungo River deposit in the eastern United States [98] and from the Oligocene deposit of Baja California [99] exhibit Ce anomalies that are similar to the Peru Shelf deposit. The anomaly for the Cretaceous deposit in the Paris Basin [89] has a slightly more negative anomaly. All of these deposits likely derived their REE from the seawater at the time of their accumulation.

According to the hypothesis of Wright et al. [83], the Ce anomaly of the Phosphoria Formation and Lisburne Group (**Figure 11**(a)) recorded an elevated ocean oxic event in the late Paleozoic era. Indeed, Berner [100] advanced an O_2 model for the atmosphere during the Phanerozoic eon (**Figure 11**(b)) that identified a significantly elevated O_2 concentration in the late Paleozoic, owing possibly to the evolution of large vascular land plants. In his model, O_2decreased sharply at the end of the Paleozoic and then

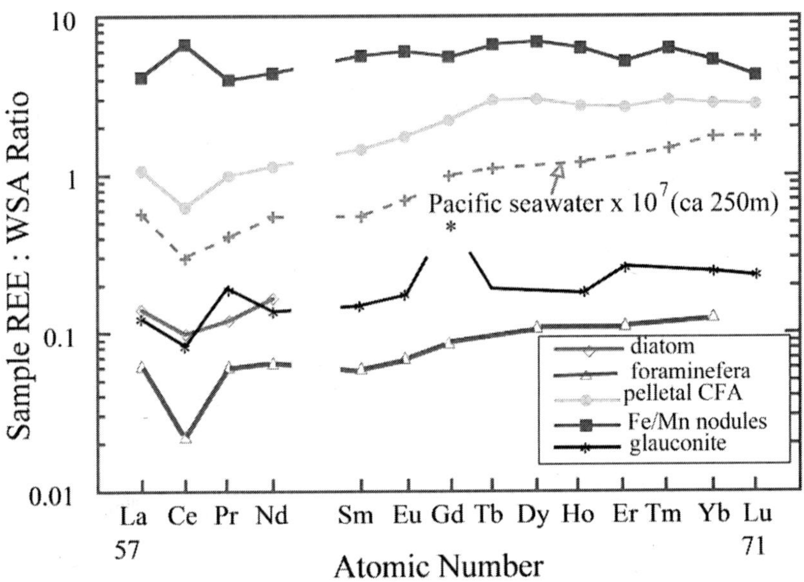

Figure 9. The REE curves of Pacific Ocean abyssal ferromanganese nodules [6,14]; foraminifera [21] and diatoms from the open ocean [6]; pelletal CFA and glauconite from 300 m depth on the Peru Shelf [19], the depth of the core of the OMZ in the eastern Pacific Ocean; and seawater from the OMZ [5]. The Gd value in glauconite may reflect analytical error.

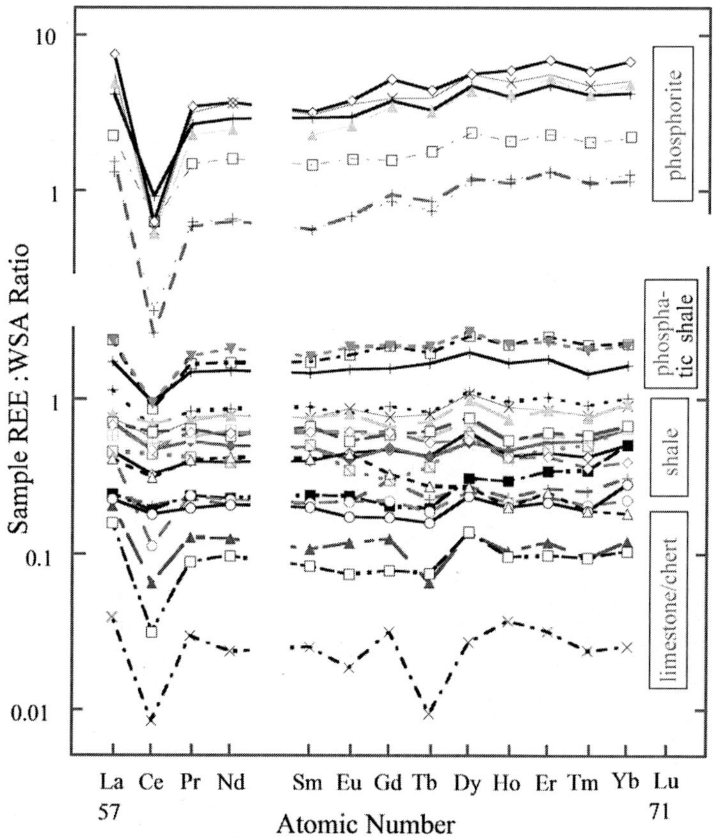

Figure 10. The REE curves of the phosphate deposit at the Enoch Valley Phosphate Mine in southeast Idaho [80]. The curves are separated into four sediment lithologies listed at the far right. Phosphorite has greater than 25% carbonatefluorapatite (CFA); phosphatic shale has 5% to 25% CFA; shale has greater than approximately 60% siliciclastic debris; limestone/chert has greater than 75% calcite + dolomite + chert.

gradually increased to reach its present value approximately 5.0×10^7 yrs ago. The Ce anomaly of pelletal CFA deposits during the late Paleozoic, Cretaceous, and Tertiary times conforms reasonably well to the O_2 atmospheric model. The deposit in the Fernie Formation, Canada, of Jurassic Age [101] clearly does not fit the atmospheric O_2 trend. Unfortunately, the texture for this deposit is poorly documented. Although the relationship between the Ce anomaly of the few Phanerozoic pelletal CFA deposits that have been examined in detail and the level of atmospheric O_2 during this geologic inteval is promising, confirmation of this relationship must await the analysis of additional carefully documented pelletal CFA deposits throughout Phanerozoic time.

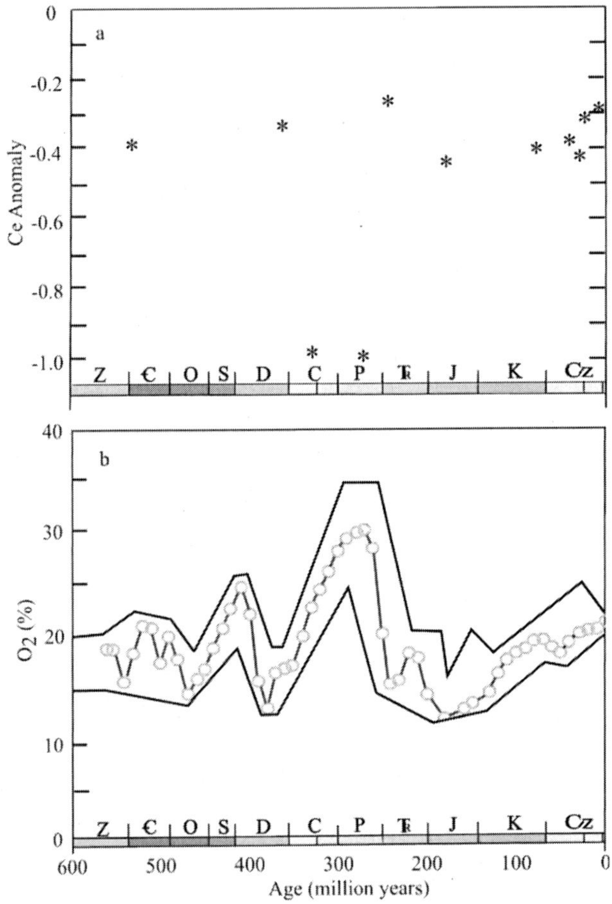

Figure 11. (a) The Ce anomaly of pelletal phosphate deposits versus geologic age (see text for references). The anomaly is defined as $\log[(3Ce_n)/(2La_n + Nd_n)]$, for which the nsub-script is the WSA-normalized value; (b) Atmospheric O_2 concentration during the Phanerozoic [100]. The symbols along the x-axis of each frame identify the geologic periods (e.g., C = Carboniferous).

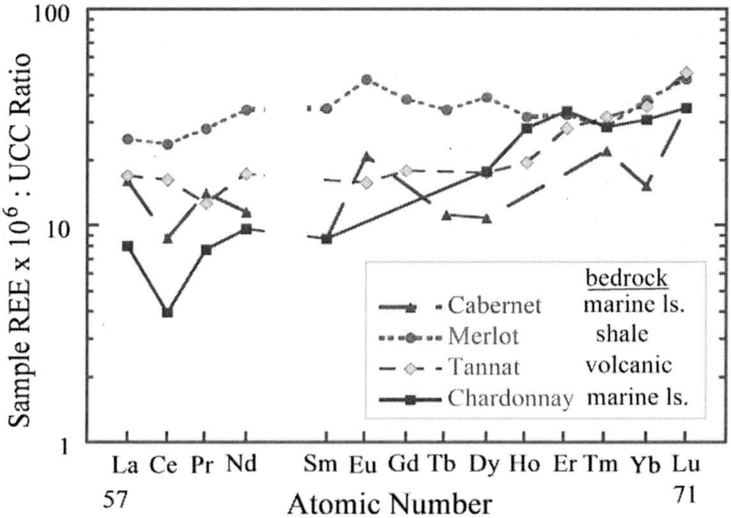

Figure 12. The REE curves of wine from South America [25]. The designation of the curves is based on the similatity to known REE curves of rock lithologies, not on the curves of the soils and bedrocks on which the grapes may actually have been grown.

6. REE IN WINE

Bentlin et al. [25] analyzed four varietal wines of South America, each of which has a unique REE curve (**Figure 12**). Each curve possibly represents the curve of the soil on which the grapes were grown and the source rock of that soil. The curves suggest that any chemical weathering that contributed to soil formation fractionated insignificantly Ce from the 3+ valence REE, i.e., between rock, ground water and soil. This would support the interpretation of limited fractionation of the REE within terrigenous siliciclastic debris, as suggested by its composition in the suspended fraction of the world's major rivers, but is at odds with the composition of the dissolved load of rivers.

As a possible alternative, the REE may have been acquired during the filtration and storage of the wines [102]. Which of these sources might determine the REE content of a wine should be easily determined by analyzing the soil and ground water on which the grapes were grown versus the various materials used in the production of the wine and possible changes to its composition during storage. If these curves do indeed reflect those of the soils, then an analysis of them should provide added evidence of the usefulness of the REE in identifying provenance.

7. CONCLUSIONS

1) The REE are a strongly coherent group of elements, whose concentrations in rocks, soils, surface waters, and sedimentary deposits conform to the Oddo-Harkins Rule, whereby even-numbered elements are more abundant than their immediate odd-numbered neighbors. The herringbone curve of their measured concentrations, when plotted against their atomic number, is converted to a smooth curve by normalizing the REE on an element-byelement basis to one of several rock standards. The standard used is unimportant except that its interelement REE concentrations should be similar to those of the samples being examined.

2) The REE in the suspended phases of seawater are partitioned into 1) a siliciclastic fraction of terrigenous origin that has a normalized REE curve similar to the curve of the average of the continental curst and 2) a seawater-derived metal oxide fraction. The REE curve of the metal-oxide fraction exhibits a negative anomaly in the surface water that becomes positive below a few meters depth. The dissolved REE fraction of seawater, in general, exhibits a negative anomaly in surface seawater that increases in its negativity with ocean depth. The dissolved fraction also exhibits an enrichment of the heavy REE relative to the light REE. It too increases with depth.

3) The REE that accumulate on the seafloor are partitioned into spatially varying amounts of a terrestrialderived lithogenous fraction, a biogenous fraction composed largely of the tests of plankton, carbonate-fluorapatite (CFA), montmorillonite (smectite), the zeolite phillipsite, glauconite, and an amorphous Fe and Mn oxihydride fraction. The lithgenous fraction has a REE curve similar to that of the terrigenous-derived fraction in the suspended load of seawater, whereas the seawaterderived fractions have one of two curves. The curve of ferromanganese oxides, in the form of abyssal

nodules and seamount crusts, is characterized by a strongly positive Ce anomaly that is similar to the curve of the aceticacid-soluble suspended phase of the water column at depths greater than several meters. By contrast, the curves for pelletal CFA and glauconite, whose accumulation is limited largely to the continental shelf and slope under seawater conditions of suboxic bacterial respiration, duplicate the curve of the immediately overlying seawater. The biogenous fraction also exhibits a negative Ce.

4) The strongly similar REE curve of the pelletal CFA that is currently accumulating on the Peru Shelf and the curve of the overlying seawater suggests that the curve of ancient pelletal CFA deposits may parallel the curve of seawater at their time and location of accumulation. As the carbon and oxygen isotopic signals of these deposits are largely retained during burial, so might also their REE curve. In support of this interpretation, pelletal CFA deposits that accumulated in Late Paleozoic time, during an interval of earth's history of elevated atmospheric O_2, have the strongest negative Ce anomaly of deposits analyzed. Pelletal CFA deposits of Tertiary age from several parts of the world exhibit REE curves that closely resemble the curve of the Peru Shelf deposit. They accumulated when the atmospheric O_2 concentration was similar to its current concentration. Thus, the REE curves of pelletal CFA deposits from the other geologic periods may also have recorded the redox conditions of ancient atmospheres and oceans.

5) Several varietal wines from South America have REE curves that parallel those of different rock lithologies. It is possible that the relative concentrations of the REE are inherited from the soil and/or ground water whose REE curves are determined by the bedrock on which the vines are grown. These similarities are another example that demonstrates the contribution REE can make in unraveling the geochemistry of both terrestrial and marine environments.

REFERENCES

1. J. S. Goldstein and S. B. Jacobsen, "Rare Earth Elements in Rivers," Earth and Planetary Science Letters, Vol. 89, No. 1, 1988, pp. 35-47.

2. H. Elderfield, R. Upstill-Goddard and E. R. Sholkovitz, "The Rare Earth Elements in Rivers, Estuaries, and Coastal Seas and Their Significance to the Composition of Ocean Waters," Geochimica et Cosmochimica Acta, Vol. 54, No. 4, 1990, pp. 971-991.
3. E. R. Sholkovitz, "The Aquatic Chemistry of Rare Earth Elements in Rivers and Estuaries," Aquatic Chemistry, Vol. 1, No. 1, 1995, pp. 1-34.
4. E. D. Goldberg, "Mineralogy and Chemistry of Marine Sedimentation," In: F. P. Shepard, Ed., Marine Geology, Harper and Row, New York, 1963, pp. 436-466.
5. D. J. W. DeBaar, M. P. Bacon, P. G. Brewer and K. W. Bruland, "Rare Earth Elements in the Pacific and Atlantic Oceans," Geochimica et Cosmochimica Acta, Vol. 49, No. 9, 1985, pp. 1943-1959.
6. H. Elderfield, C. J. Hawkesworth, M. J. Greaves and S. E. Calvert, "Rare Earth Element Geochemistry of Oceanic Ferromanganese Nodules and Associated Sediments," Geochimica et Cosmochimica Acta, Vol. 45, No. 4, 1981, pp. 513-528.
7. E. R. Sholkovitz and D. L. Schneider, "Cerium Redox Cycles and Rare Earth Elements in the Sargasso Sea," Geochimica et Cosmochimica Acta, Vol. 55, No. 10, 1991, pp. 2737-2743.
8. D. J. Piepgras and S. B. Jacobsen, "The Behavior of Rare Earth Elements in Seawater: Precise Determination of Variations in the North Pacific Water Column," Geochimica et Cosmochimica Acta, Vol. 56, No. 5, 1992, pp. 1851-1862.
9. S. R. Sholkovitz, W. M. Landing and B. L. Lewis, "Ocean Particle Chemistry: The Fractionation of Rare Earth Elements between Suspended Particles and Seawater," Geochimica et Cosmochimica Acta, Vol. 58, No. 6, 1994, pp. 1567-1580.
10. J. H. Lee and R. H. Byrne, "Complexation of Trivalent Rare Earth Elements (Ce, Eu, Gd, Tb, Yb) by Carbonate," Geochimica et Cosmochimica Acta, Vol. 57, No. 2, 1993, pp. 295-302.

11. R. H. Byrne, "Inorganic Speciation of Dissolved Elements in Seawater," Geochemical Transactions, Vol. 3, No. 2, 2002, pp. 11-16.

12. Y. R. Luo and R. H. Byrne, "Carbonate Complexation of Yttrium and the Rare Earth Elements in Natural Waters," Geochimica et Cosmochimica Acta, Vol. 68, No. 4, 2004, pp. 691-699.

13. R. H. Byrne and K. H. Kim, "Rare-Earth Element Scavenging in Seawater," Geochimica et Cosmochimica Acta, Vol. 54, No. 10, 1990, pp. 2645-2656.

14. D. Z. Piper, "The Metal Oxide Fraction of Pelagic Sediment in the Equatorial North Pacific Ocean: A Source of Metals in Ferromanganese Nodules," Geochimica et Cosmochimica Acta, Vol. 52, No. 8, 1988, pp. 2127-2145.

15. P. Halbach and D. Puteanus, "The Influence of the Carbonate Dissolution Rate on the Growth and Composition of Co-Rich Ferromanganese Crusts from Central Pacific Seamount Areas," Earth and Planetary Science Letters, Vol. 68, No. 1, 1984, pp. 73-87.

16. M. Bau, A. Koschinsky, P. Dulski and J. R. Hein, "Comparison of the Partitioning Behaviors of Yttrium, RareEarth Elements, and Titanium between Hydrogenetic Marine Ferromanganese Crusts and Seawater," Geochimica et Cosmochimica Acta, Vol. 60, No. 10, 1996, pp. 1709- 1725.

17. J. R. Hein, A. Koschinsky, P. Halbach, F. T. Manheim, M. Bau, J. K. Kang and N. Lubick, "Iron and Manganese Oxide Mineralizations in the Pacific," Geological Society, London, Special Publication, Vol. 119, No. 1, 1997, pp. 123-138.

18. M. Bau and A. Koschinsky, "Hafnium and Neodymium Isotopes in Seawater and Ferromanganese Crusts: The Element Perspective," Earth and Planetary Science Letters, Vol. 241, No. 3-4, 2006, pp. 952-961.

19. D. Z. Piper, P. A. Baedecker, J. G. Crock, W. C. Burnett and B. L. Loebner, "Rare Earth Elements in the Phosphatic-Enriched Sediments of the Peru Shelf," Marine Geology, Vol. 80, No. 3-4, 1988, pp. 269-285.

20. D. Z. Piper, "Rare-Earth Elements in the Sedimentary Cycle: A Summary," Chemical Geology, Vol. 14, No. 4, 1974, pp. 285-304.

21. P. A. Palmer, "Rare Earth Elements in Foraminifera Tests," Earth and Planetary Science Letters, Vol. 73, No. 2-4, 1985, pp. 285-298.

22. F. Guichard, T. M. Church, M. Treull and H. Jaffrezic, "Rare Earth Elements in Barites: Distribution and Effects on Aqueous Partitioning," Geochimica et Cosmochimica Acta, Vol. 43, No. 7, 1979, pp. 983-997.

23. L. Haskin and M. A. Gehl, "The Rare-Earth Distribution in Sediments," Journal of Geophysical Research, Vol. 67, No. 6, 1962, pp. 2537-2541.

24. Ø. Y. Høgdahl, B. T. Bowen and S. Melson, "Neutron Activation Analysis of Lanthanide Elements in Seawater," Advances in Chemistry Series, Vol. 73, 1968, pp. 308-325.

25. F. R. S. Bentlin, C. M. M. dos Santos, E. M. M. Flores and D. Pozebon, "Lanthanides Determination in Red Wine Using Ultrasound Assisted Extraction, Flow Injection, Aerosol Disolvation and ICP-MS," Analytica Chimica Acta, Vol. 710, 2012, pp. 33-39.

26. R. A. Schmidt, R. H. Smith, J. E. Lasch, A. W. Mosen, D. A. Olehy and J. Vasilevshis, "Abundances of Fourteen Rare-Earth Elements, Scandium, and Yttrium in Meteoritic and Terrigenous Matter," Geochimica et Cosmochimica Acta, Vol. 27, No. 6, 1963, pp. 577-622.

27. T. R. Wildman and L. Haskin, "Rare-Earth Elements in Ocean Sediments," Journal of Geophysical Research, Vol. 70, No. 12, 1965, pp. 2905-2911.

28. B. Mason, "Principals of Geochemistry," John Wiley and Sons, New York, 1966.

29. M. A. Haskin and L. Haskin, "Rare Earth Elements in European Shales, a Redetermination," Science, Vol. 154, No. 3748, 1964, pp. 507-509.

30. W. E. Wedepohl, "The Composition of the Continental Crust," Geochimica et Cosmochimica Acta, Vol. 59, No. 7, 1995, pp. 1217-1232.

31. P. L. Gromet, P. F. Dymek, L. A. Haskin and R. L. Korotev, "The North American Shale Composite: Its Composition, Major and Minor Element Characteristics," Geochimica et Cosmochimica Acta, Vol. 48, No. 12, 1984, pp. 2469-2482.

32. K. H. Wedepohl, "Handbook of Geochemistry, Volume I-IV," Springer-Verlag, Berlin, 1969-1978.

33. R. L. Rudnick and S. Gao, "Composition of the Continental Crust," In: H. D. Holland and K. K. Turekian, Eds., Treatise on Geochemistry, Elsevier, New York, 2003, pp. 1-64.

34. S. M. McLennan, "Rare Earth Elements in Sedimentary Rocks: Influence of Provenance and Sedimentary Processes," Reviews in Mineralogy and Geochemistry, Vol. 21, No. 1, 1989, pp. 169-200.

35. S. M. McLennan, "Relationships between the Trace Element Composition of Sedimentary Rocks and Upper Continental Crust," Geochemistry Geophysics Geosystems, Vol. 2, No. 4, 2001, pp. 1-24.

36. Ø. Høgdahl, "Distribution of Rare Earth Elements in Seawater," NATO Research Grant 203, Semiannual, Progress Report No. 2-6, 1965-1967.

37. J. Gaillardet, J. Viers and B. Dupré, "Trace Elements in River Waters," In: J. I. Drever, Ed., Treatise on Geochemistry, Surface and Groundwater, Weathering and Soils, Elsevier, Amsterdam, 2003, pp. 225-272.

38. D. Z. Piper, S. Ludington, D. S. Duval and H. E. Taylor, "Geochemistry of Bed and Suspended Sediment in the Mississippi River System: Provenance Versus Weathering and Winnowing," Science of the Total Environment, Vol. 362, No. 1-3, 2005, pp. 179-204.

39. C. R. German and H. Elderfield, "Rare Earth Elements in the NW Indian Ocean," Geochimica et Cosmochimica Acta, Vol. 54, No. 7, 1990, pp. 1029-1040.

40. H. Elderfield and M. J. Greaves, "The Rare Earth Elements in Seawater," Nature, Vol. 296, No. 5854, 1982, pp. 214-219.

41. H. U. Sverdrup, M. W. Johnson and R. H. Fleming, "The Oceans—Their Physics, Chemistry, and General Biology," Prentice-Hall, Upper Saddle River, 1942.

42. K. W. Bruland, "Trace Elements in Seawater," In: J. P. Riley and R. Chester, Eds., Chemical Oceanography, Academic Press, London, 1983, pp. 158-220.

43. W. M. Latimer, "Oxidation Potentials," Prentice Hall, New York, 1953.

44. D. D. Wagman, W. H. Evans, V. B. Parker, R. H. Schumm, I. Halow, S. M. Bailey, K. L. Churney and R. L. Nutall, "The NBS Tables of Chemical Thermodynamic Properties: Reference Data," Journal of Physical Chemistry, Vol. 11, Supplement 2, 1982.

45. W. M. Landing and K. W. Bruland, "The Biochemistry of Iron and Manganese in the Pacific Ocean," Geochimica et Cosmochimica Acta, Vol. 51, No. 1, 1987, pp. 29-43.

46. W. S. Broecker and T. H. Peng, "Tracers in the Sea," Eldigio Press, Palisades, New York, 1982.

47. J. W. Murray, B. Spell and B. Paul, "The Contrasting Geochemistry of Manganese and Chromium in the Eastern Tropical Pacific Ocean," In: C. S. Wong, E. Boyle, K. W. Bruland, J. D. Burton and E. D. Goldberg, Eds., Trace Metals in Seawater, Plenum, New York, 1983, pp. 643- 669.

48. H. Elderfield, "Chromium Speciation in Seawater," Earth and Planetary Science Letters, Vol. 9, No. 1, 1970, pp. 10-16.

49. C. I. Measures, R. E. McDuff and J. M. Edmond, "Selenium Redox Chemistry at GEOSECS I Reoccupation," Earth and Planetary Science Letters, Vol. 49, No. 1, 1980, pp. 102-108.

50. G. A. Cutter and K. W. Bruland, "The Marine Biogeochemistry of Se: A Reevaluation," Limnology and Oceanography, Vol. 29, No. 6, 1984, pp. 1179-1192.

51. D. Langmuir, "Uranium Solution-Mineral Equilibria at Low Temperatures with Applications to Sedimentary Ore

Deposits," Geochimica et Cosmochimica Acta, Vol. 42, No. 6, 1978, pp. 547-569.

52. R. W. Collier, "Particulate and Dissolved Vanadium in the North Pacific Ocean," Nature, Vol. 309, No. 5967, 1984, pp. 441-444.

53. L. Jacobs, S. R. Emerson and J. Skei, "Partitioning and Transport of Metals across the O_2/H_2S Interface in a Permanently Anoxic Basin, Framvaren Fjord, Norway," Geochimica et Cosmochimica Acta, Vol. 49, No. 6, 1985, pp. 1433-1444.

54. E. A. Boyle, F. R. Sclater and J. M. Edmond, "On the Marine Geochemistry of Cadmium," Nature, Vol. 263, No. 5572, 1976, pp. 42-44.

55. R. W. Collier, "Molybdenum in the Northeast Pacific Ocean," Limnology and Oceanography, Vol. 30, No. 6, 1985, pp. 1351-1354.

56. S. R. Emerson and S. S. Huested, "Ocean Anoxia and the Concentrations of Molybdenum and Vanadium in Seawater," Marine Chemistry, Vol. 34, No. 3-4, 1991, pp. 177-196.

57. Y. Nozaki, "Rare Earth Elements and Their Isotopes in the Ocean," Encyclopedia of Ocean Science, Academic Press, 2001, pp. 2354-2366.

58. J. W. Moffett, "Micobially Mediated Cerium Oxidation in Seawater," Nature, Vol. 345, No. 6274, 1990, pp. 421- 423.

59. S. R. Taylor, S. M. McLennan and M. T. McCulloch, "Geochemistry of Loess, Continental Crustal Composition and Crustal Model Ages," Geochimica et Cosmochimica Acta, Vol. 47, No. 11, 1983, pp. 1897-1905.

60. L. L. Feng, Z. G. Hu, J. T. Ju and L. P. Zhu, "Variations in Trace Elements (Including Rare Earth Elements) Concentrations with Grain Sizes in Loess and Their Implications for Tracing the Provenance of Eolian Deposits," Quaternary International, Vol. 236, No. 1-2, 2011, pp. 116-126.

61. L. Moa, D. Mo, M. Li, K. Zhou, J. Yang and W. Guo, "The Rare Earth Element Composition of Sediments from the Loess Tableland in the Liyang Plain, Southern China: Implications

for Provenance and Weathering Intensity," Environmental Earth Science, Vol. 62, No. 8, 2011, pp. 1609-1617

62. M. Ferrat, D. J. Weiss, et al., "Improved Provenance Tracing of Asian Dust Sources Using Rare Earth Elements and Selected Trace Elements for Palaeomonsoon Studies on the Eastern Tibetan Plateau," Geochimica et Cosmochimica Acta, Vol. 75, No. 21, 2011, pp. 6374- 6399.

63. V. M. Goldschmidt, "Geochemistry," Oxford University Press, Oxford, 1954.

64. S. E. Calvert and T. F. Pedersen, "Elemental Proxies for Palaeoclimatic and Palaeoceanographic Variability in Marine Sediments: Interpretation and Application," In: C. Hillaire-Marcel and A. de Vernal, Eds., Paleoceanography of the Late Cenozoic, Part 1, Methods, Elsevier, New York, 2007, pp. 567-644.

65. D. Z. Piper and S. E. Calvert, "Holocene and Late Glacial Palaeoceanography and Palaeolimnology of the Black Sea: Changing Provenance and Basin Hydrography over the Past 25,000 Years," Geochimica et Cosmochimica Acta, Vol. 75, No. 19, 2011, pp. 5597-5624.

66. W. M. Landing and B. L. Lewis, "Thermodynamic Modeling of Trace Metal Speciation in the Black Sea," In: J. W. Murray and E. Izdar, Eds., Black Sea Oceanography, Kluwer, Dordrecht, 1991, pp. 125-160.

67. H. J. W. DeBaar, C. R. German, H. Elderfield and P. van Gaans, "Rare Earth Element Distributions in Anoxic Waters of the Cariaco Trench," Geochimica et Cosmochimica Acta, Vol. 52, No. 5, 1988, pp. 1203-1219.

68. J. Schijf, H. J. W. DeBaar, J. R. Wijbrands and W. M. Landing, "Dissolved Rare Earth Elements in the Black Sea," Deep-Sea Research, Vol. 38, Supplement 2, 1991, pp. S805-S823.

69. C. R. German, B. P. Holliday and H. Elderfield, "Redox Cycling of Rare Earth Elements in the Suboxic Zone of the Black Sea," Geochimica et Cosmochimica Acta, Vol. 55, No. 12, 1991, pp. 3553-3558.

70. R. L. Cullers, "The Chemical Signature of Source Rocks in Size Fractions of Holocene Stream Sediment Derived from Metamorphic Rocks in the Wet Mountains Region, USA," Chemical Geology, Vol. 113, No. 3-4, 1988, pp. 327-343.

71. R. L. Cullers and V. N. Podkovyrov, "Geochemistry of the Mesoproterozoic Lakhanda Shales in Southeastern Yakutia, Russia: Implications for Mineralogical and Provenance Control Recycling," Pre-Cambrian Research, Vol. 104, No. 1-2, 2000, pp. 77-93.

72. B. Yan, W. Yan, L. Miao, L. Huang and Z. Chen, "Geochemical Characteristics and Provenance Implications of Rare Earth Elements in Surface Sediments from Bays along Guangdong Coast, Southeast China," Environmental Earth Science, Vol. 65, 2010, pp. 2195-2205.

73. F. Xu, A. Li, T. Li, K. Xu, S. Chen, L. Qiu and Y. Cao, "Rare Earth Element Geochemistry in the Inner Shelf of the East China Sea and Its Implication for Sediment Provenances," Journal of the Rare Earths, Vol. 29, No. 7, 2011, pp. 702-709.

74. G. Müller and P. Stoffers, "Mineralogy and Petrology of Black Sea Basin Sediments," In: E. T. Degens and R. A. Ross, Eds., The Black Sea—Geology, Chemistry and Biology, American Association of Petroleum Geologists: Memoir, American Association of Petroleum Geologists, Tulsa, 1974, pp. 200-248.

75. O. Yigiterhan and J. W. Murray, "Trace Metal Composition of Particulate Matter of the Danube River and Turkish Rivers Draining into the Black Sea," Marine Chemistry, Vol. 111, No. 1-2, 2008, pp. 63-76.

76. E. T. Degens and D. A. Ross, "The Black Sea—Geology, Chemistry and Biology," In: E. T. Degens and D. A. Ross, Eds., American Association of Petroleum Geologists: Memoir, American Association of Petroleum Geologists, Tulsa, 1974, 633 p.

77. O. Kwiecien, H. W. Arz, F. Lamy, B. Plessen, A. Bahr and G. H. Haug, "North Atlantic Control on Precipitation Curves in the Eastern Mediterranean/Black Sea Region during the Last

Glacial," Quaternary Research, Vol. 71, No. 3, 2009, pp. 375-384.

78. E. D. Goldberg, M. Koide, R. A. Schmitt and R. M. Smith, "Rare Earth Distributions in the Marine Environment," Journal of Geophysical Research, Vol. 68, No. 14, 1963, pp. 4209-4217.

79. H. Elderfield, "The Oceanic Chemistry of the Rare-Earth Elements," Philosophical Transaction of the Royal Society of London, A, Vol. 325, No. 1583, 1988, pp. 105-126.

80. D. Z. Piper, "Marine Chemistry of the Permian Phosphoria Formation and Basin, Southeast Idaho," Economic Geology, Vol. 96, No. 3, 2001, pp. 599-620.

81. F. A. Richards, "Coastal Upwelling," In: F. A. Richards, Ed., Coastal Estuarine Science, American Geophysical Union, Washington DC, Vol. 1, 1981.

82. J. R. Hein, C. L. Fleishman, L. A. Morgenson, S. H. Bloomer and R. A. Stern, "Submarine Ferromanganese Deposits from the Mariana and Volcano Volcanic Arcs, West Pacific," US Geological Survey Open File Report 87-281, 1987.

83. J. Wright, H. Schrader and W. T. Holser, "Paleoredox and Variations in Ancient Oceans Recorded by Rare Earth Elements in Fossil Apatite," Geochimica et Cosmochimica Acta, Vol. 51, No. 3, 1987, pp. 631-644.

84. G. J. Demaison and G. T. Moore, "Anoxic Environments and Oil Source Bed Genesis," American Association of Petroleum Geologists Bulletin, Vol. 64, No. 8, 1980, pp. 1179-1209.

85. T. Hallam, "Catastrophes and Lesser Calamities: The Causes of Mass Extinctions," Oxford University Press, Oxford, 2004.

86. K. M. Meyer and L. R. Kump, "Oceanic Euxinia in Earth History: Causes and Consequences," Annual Reviews of Earth and Planetary Science Letters, Vol. 36, No. 1, 2008, pp. 251-288.

87. H. Jenkyns, "Geochemistry of Ocean Anoxic Events," Geochemistry Geophysics Geosystems, Vol. 11, No. 3, 2010, pp. 1-30.

88. W. T. Holser, "Evaluation of the Application of RareEarth Elements to Paleoceanography," Paleoceanography, Vol. 132, 1997, pp. 309-323.

89. I. Jarvis, "Rare Earth Element Geochemistry of Late Cretaceous Chalks and Phosphorites from Northern France," Geological Survey of India Special Publication, Vol. 17, 1984, pp. 179-190.

90. H. Elderfield and R. Pagett, "Rare Earth Elements in Ichthyoliths: Variations with Redox Conditions and Depositional Environments," Science of the Total Environment, Vol. 49, 1986, pp. 175-197.

91. P. Grandjean-Lécuyer, R. Feist and F. Albarède, "Rare Earth Elements in Old Biogenic Apatite," Geochimica et Cosmochimica Acta, Vol. 57, No. 11, 1993, pp. 2507- 2514.

92. G. Shields and P. Webb, "Diagenetic Constraints on the Use of Cerium Anomalies as Palaeoseawater Redox Proxies: An Isotopic and REE Study of Cambrian Phosphorites," Chemical Geology, Vol. 175, No. 1-2, 2001, pp. 29-48.

93. L. Kocsis, C. N. Treuman and M. R. Palmer, "Protracted Diagenetic Alteration of REE Contents in Fossil Bioapatites: Direct Evidence from Lu-Hf Isotope Systematics," Geochimica et Cosmochimica Acta, Vol. 74, No. 21, 2010, pp. 6077-6092.

94. C. R. German and H. Elderfield, "Application of the Ce Anomaly as a PaleoredoxIndicatior: The Ground Rules," Paleoceanography, Vol. 5, No. 5, 1990, pp. 823-833.

95. A. Shemesh, Y. Kolodny and B. Luz, "Isotope Geochemistry of Oxygen and Carbon in Phosphate and Carbonate of Phosphorite Francolite," Geochimica et Cosmochimica Acta, Vol. 52, No. 11, 1988, pp. 2565-2572.

96. Y. Kolodny and B. Luz, "Phosphate Deposits, Formation and Diagenetic History," In: N. Clauer and S. Chaudhuri, Eds., Isotopic Signatures and Sedimentary Records, Springer Verlag, New York/Heidelberg, 1984, pp. 69- 122.

97. J. A. Dumoulin, J. F. Slack, M. T. Whalen and A. G. Harris, "Depositional Setting and Geochemistry of Phosphorites and Metalliferous Black Shales in the Carboniferous-Permian

Lisburne Group, Northern Alaska," US Geological Survey Professional Paper 1776, 2011.

98. S. Riggs, S. W. Snyder, D. Ames and P. Stille, "Chronostratigraphy of Upper Cenozoic Phosphorites on the North Carolina Continental Margin and the Oceanographic Implications for Phosphogenesis," Society of Economic Paleontologists and Mineralogists Special Publication, Vol. 66, 2000, pp. 369-385.

99. D. Z. Piper, "Geochemistry of a Tertiary Sedimentary Phosphate Deposit: Baja California Sur," Chemical Geology, Vol. 92, No. 4, 1991, pp. 283-316.

100. R. A. Berner, "GEOCARBSULF: A Model for Phanerozoic Atmospheric O_2 and CO_2," Geochimica et Cosmochimica Acta, Vol. 70, No. 23, 2006, pp. 5653-5664.

101. D. E. MacDonald, "Geology and Resource Potential of Phosphates in Alberta and Portions of Southwest British Columbia," M.Sc. Thesis, University of British Columbia, Vancouver, 1985.

102. N. Jakubowski, R. Brandt, D. Stuewer, H. R. Eschnauer and S. Görtges, "Analysis of Wines by ICP-MS: Is the Curve of Rare Earth Elements a Reliable Fingerprint for the Provenance," Fresenius Journal of Analytical Chemistry, Vol. 364, No. 5, 1999, pp. 424-428.

8

CHAPTER

Separation of Kaolinite from Ion-Adsorption Rare Earth Tailings in Southern China and Iron Removal Treatment

Yongqing Wang, Huayin Liang, Qibing Chang*, Xiaozhen Zhang, Jian'er Zhou

School of Materials Science and Engineering, Jingdezhen Ceramic Institute, Xianghu, Jingdezhen, China

ABSTRACT

Several hundred million tons of ion-adsorption rare earth tailings exist in Ganzhou, Southern China, which is a severe environmental hazard. To reduce and reutilize the tailing, kaolinite has been separated from the tailings by mechanical separation in laboratory scale and pilot scale. The results show that the tailing is mainly composed of fine kaolinite and coarse quart. Quartz and kaolinite can be separated by sieves, shaker, spiral chute or hydrocyclone, which has the similar results in laboratory scale and pilot scale. 30.2% of the tailings can be re-sourced and applied in ceramic industries. 41.7% of kaolinite can be obtained after sorting and iron removal by magnetic separator in pilot scale, which can be applied in ceramic industries according to the Chinese national standard (TC-3). The results give a progressive solution to re-source the tailings economically.

Keywords: Rare Earth Tailing, Kaolinite, Separation, Reutilize, Iron Removal

1. INTRODUCTION

Rare earth elements (REEs) have become vital and indispensable components of many high-tech products, devices and technologies. China's production of rare earth minerals has provided more than 90% of the world's supply since 2001 (Su, 2009; Chen, 2011) [1] [2] . However, what is left in China is million tons of tailings because rare earth ore deposits have relatively low REEs concentrations ranging from 10 to a few hundred parts per million by weight (Yang, et al., 2013) [3] . Ion-adsorption rare earth reserves only distribute in seven provinces of southern China (Su, 2009; Wübbeke, 2013) [1] [4] . The ion-adsorption rare earth deposits were formed by chemical weathering decomposition and dissolution of granite and granite porphyry (containing relative high abundance of rare earth elements) and subsequent adsorption and enrichment on clay minerals during the migration and penetration process of rare earth mineral solutions. Unlike other rare earth minerals which are in solid state mineral phase and tend to be symbiotic or associated with the radioactive elements uranium and thorium, ion-adsorption rare earth minerals occur at a simple trivalent cationic state which is simply adsorbed onto clay sand and can be readily extracted by a simple leaching technique via anion-exchange process (Yang, et al., 2013) [3] . Hence, the tailings can be and must be re-sourced because there are 191 million tons of tailings just distributed in Ganzhou region (Jiangxi Province, Southern China).

Kaolinite and quartz are the main components of the tailings characterized by Mineral Liberation Analyzer. It is possible to separate of kaolinite from the tailings due to the differences of density and particle size of kaolinite and quartz. In the present work, several methods of mineral separation were applied to the tailings and a systemic separation of the tailing was proposed.

2. MATERIALS AND METHODS

2.1. Collection and Pre-Treatment of Tailings Samples

The tailings were collected from the ores of ion-adsorption rare earths in Ganzhou, Southern China. Representative sample was made by following Chinese standard method of mixing (China Standard Specification, GB 2007. 1-87). No further grinding treatment was conducted in laboratory. The tailings were dried in oven at 100°C overnight, cooled naturally, and then be characterized.

The tailings was suspended in water (1:4 tailings to water ratio) and slurried by stirring with an impeller stirrer without adding any dispersant or adjusting the pH value.

2.2. Characterization

The bulk chemical element analysis in samples melted with lithium tetraborate was performed with an XRF spectrometer (Axios mAX PANalytical, Netherland).

The bulk mineral composition was performed with Mineral Liberation Analyzer (MLA 650, Czech Republic).

The particle size distribution of the tailings was measured by wet sieving method using 20 wt% suspensions. The sieve size of the copper standard sieves were sequentially 0.8 mm, 0.4 mm, 0.25 mm, 0.12 mm, 0.075 mm, and 0.043 mm.

2.3. Sorting Experiments in Laboratory Scale

Sorting of tailing by shaker: 500 ml 20 wt% tailings slurry was put into a shaker (6-S, China). The separation condition kept that: the stroke is 10 mm; the frequency is 220/min.

Sorting of tailing by settlement: 500 ml 20 wt% tailings slurry was sieved by wet sieving. The −0.043 mm fraction of tailings was further slurried in water (about 10 wt% solid) and then put into a cylinder (500 ml in volume). The slurry is stood undisturbed for 1 hr. The height of the slurry was separated according to Stoke equation of 5 μm, 10 μm and 20 μm.

Iron removal: 500 ml 20 wt% tailings slurry was sieved by wet sieving. The −0.043 mm fraction of tailings was put into a cycle type high gradient magnetic separator (SSS-I-145) (Chen, L., 2011, Svoboda, J., 2003) [5] [6] . The magnetic field intensity is in the range of 0.2 T-1.0T.

For the three processes above-mentioned, all the classified tailings were collected, and then dried in oven at 100°C overnight. The dried tailings were weight and characterized by XRF.

2.4. Separation Experiments in Pilot Scale

200 kg tailings were suspended in water to form 20 wt% slurry. The flow chart of the separation of the slurry is shown in Figure 1. The slurry was firstly sieved by a high frequency vibrating sieve (GYX11-1210, sieve size is 0.4 mm). The −0.44 mm fraction of the slurry was delivered into a spiral chute (TGL 0610) by a high-pressure pump. The K4, K5 fraction of the slurry were separated by a hydrocyclone. The overflow was separated by the hydrocyclone secondly. Finally, the second overflow was put into wet dual band, high intensity, and high gradient magnetic separator (SSS-I-800). The magnetic field intensity is in the range of 0.6 T.

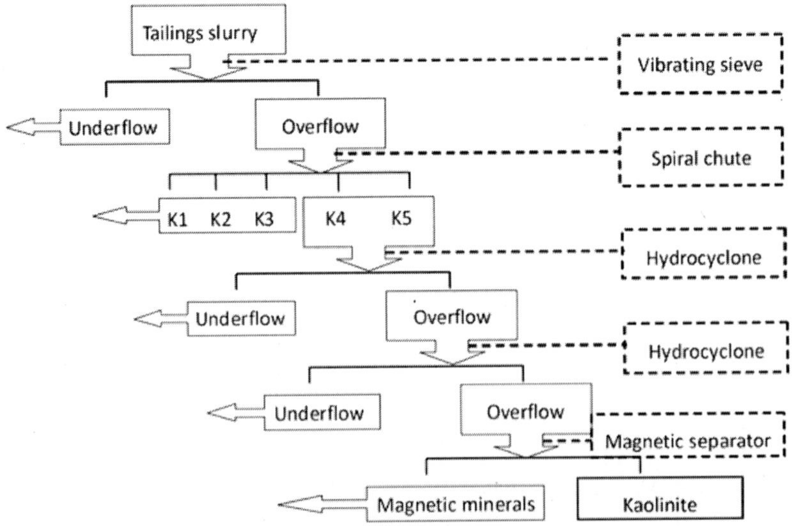

Figure 1. Flow chart of the tailings separation in pilot scale.

For the processes above-mentioned, all the classified tailings were collected, and then dried in oven at 100°C overnight. The dried tailings were weight and characterized by XRF.

3. RESULTS AND DISCUSSIONS

3.1. Characterization of the Tailings

3.1.1. Chemical Composition of Tailings by MLA

Table 1 shows the mineral composition of the tailings. As can be seen, kaolinite and Quartz is the major mineral of the tailings, which shares 91.98% of the tailings. It implies that the tailings can be re-utilized in ceramic application through the separation of kaolinite or quartz. The other silicates are feldspar, biotite, muscovite and amphibole. These silicates do not need the separation before the ceramic application because the contents are small and they also can be used as sintering aid in the calcinations of ceramic. The iron and titanium minerals include of titanomagnetite, limonite, pyrite, ilmenite, rutile, iron sphene and leucoxene, which shares about 1.77% of the tailings. The iron and titanium minerals should be removed down to a limit concentration because they are harmful to the product quality of ceramic. The content of the rare earth mineral is very slow (0.04%), including of monazite, xenotime, parisite and colloidal rare earth. There is no radioactive minerals remained, indicating the re-source of the tailings is safe in ceramic application.

3.1.2 Chemical Composition of Tailings by XRF

The chemical composition of the tailings sample is presented in Table 2. As can be seen, the rich major chemical composition in oxides in mine tailing are SiO_2 (78.32%), Al_2O_3 (17.50%) and Fe_2O_3(2.11%). Pure kaolinite ($Al_2O_3 \cdot 2SiO_2 \cdot 2H_2O$) is white in color and its chemical composition is 46.54% SiO_2, 39.50% Al_2O_3 and 13.96% H_2O (Grimshaw, 1971) [7] . Assumed that all of Al_2O_3(17.50%) exist in kaolinite, it can be calculated that 44.3% of the tailings is kaolinite, which is similar to the results in Table 1. It is deduced that most of Al_2O_3 exist in kaolinite. Hence, the content of kaolinite can be estimated by that of Al_2O_3 in this work.

Minor quantities of transition elements such as iron, titanium and manganese are generally present as ancillary minerals which adversely affect the optical properties of kaolins, especially, Fe-containing minerals (Chandrasekhar and Ramaswamy, 2006; Malengreau et al., 1996) [8] [9] . They cause brown-yellow colorations and reduce the whiteness. The Chinese national standard of kaolin (TC-3, GB/T 14563-2008) applied in ceramic industries requires that $Al_2O_3 \geq 28\%$, $Fe_2O_3 \leq 1.8\%$, $TiO_2 \leq 0.6\%$ (mass ratio). Obviously, the Fe_2O_3 content is 2.11% is larger than the requirement, implying that iron removal is required before the utilization of the tailings. There is no further treatment to remove TiO_2 in the separation of the tailings because the content of TiO_2 is low.

The tailing is sieved by the standard sieves. The particle size distribution and the grade of SiO_2, Al_2O_3, Fe_2O_3 and TiO_2 in every distribution are listed in Table 3. As can be seen, the particle size of the tailing shows the bimodal distribution of 0.8mm and −0.043 mm. The grade of SiO_2 is larger than 91.46 for the + 0.4 mm fraction, implying that coarse grains are quartz and can be separated by 0.4 mm sieve. The grade of SiO_2 decreases and the grade of Al_2O_3 increase with the decrease of the particle size. Hence, most of kaolinite exists in the −0.043 mm fraction. Kaolinite and quartz maybe separated by sieving. However, the iron content in the −0.043 mm fraction is 2.16%, implying that the iron-bearing minerals are mixed with kaolinite. Iron removal is required for the kaolinite separated from the tailings (Svoboda, J. and Fujita, T., 2003, Santos, et al. 2012, Chandrasekhar and Ramaswamy, 2007, Xia, et al. 2012, Zegeye, A. et al., 2013) [6] [10] - [13] .

Table 1. Mineral composition of the tailings.

Mineral	Content %	Mineral	Content %	Mineral	Content %
Kaolinite	42.29	Monazite	0.01	Colloidal rare earth	0.01
Quartz	49.69	Xenotime	0.01	Titanomagnetite	1.10
Feldspar	3.90	Parisite	0.01	Limonite	0.29
Diopside	0.01	Rutile	0.09	Ilmenite	0.13
Muscovite	0.68	Leucoxene	0.10	Iron sphene	0.04
Biotite	1.36	Pyrite	0.01	Remainder	0.14
Amphibole	0.11	Galena	0.01	Total	100

Table 2. Chemical composition of the tailings measured by XRF/%.

SiO_2	Al_2O_3	Fe_2O_3	TiO_2	K_2O	Na_2O	CaO	MgO
78.32	17.50	2.11	0.38	1.21	0.43	0.22	1.33

3.2. Sorting Experiments in Laboratory Scale

3.2.1. Sorting of the Tailings by Shaker

The tailing can be sorted by the shaker due to the difference of the density (the density of kaolinite is 1.58 - 1.63 g/cm^3; the density of quartz is 2.64 - 2.71 g/cm^3). Table 4 shows the sorting results of the tailings. Obviously, the heavy quartz remains in the underflow K1and the middling K2, whose gradation is over than 90%. The abnormal data in underflow may be the results that some of kaolinites are entrained by quartz grains in large size. The gradation of Al_2O_3 increases with the decrease of the particle size of the tailing. It is worth mentioning that the gradation of Al_2O_3 is 24.46%, which is far larger than the other zone. The yield of the tailings in overflow K5 is 57.71%, implying that most of kaolinite is fine and locates in the overflow K5. Hence, the tailings can be preliminarily separated into kaolinite and quartz by shaker. The gradation of Fe_2O_3 is 1.94% in the overflow, which is still higher than the national standard of TC-3. The recovery ratio of Fe_2O_3 in overflow is 69.90%. Most of the fine Fe-bearing minerals are mixed with kaolinite. It is no use to remove iron by the shaker.

3.2.2. Sorting of the Tailings by Settlement

The gradation of Al_2O_3 in the overflow of the shaker is 24.46%, which doesn't meet the requirement of TC-3 (in which, $Al_2O_3 \geq 28\%$). The tailings need to be sorted further.

Table 5 shows the sorting results of the tailings by settlement. The gradation of Al_2O_3 is larger than 28% if the particle size of the tailings is no more than 10 μm. Kaolinite required in ceramic industries can be separated from the rare earth tailing under the favorable sorting condition.

3.2.3. Iron Removal

It doesn't require separating kaolinite from quartz completely because they are both needed in traditional ceramic industries. Kaolinite and quartz share 91.98% of the ion-adsorption rare earth tailings (see Table 1). It implies that the tailing can be re-utilized in common ceramic industries if the Fe_2O_3 content can be decreased to 1.8%.

Table 3. Particle size distribution of the tailing and the quality of SiO_2, Al_2O_3, Fe_2O_3 and TiO_2.

Particles size/mm	Yield/ %	Gradation/%				Recovery ratio/%			
		SiO_2	Al_2O_3	Fe_2O_3	TiO_2	SiO_2	Al_2O_3	Fe_2O_3	TiO_2
0.8	19.1	98.44	0.91	0.31	0.05	23.99	0.99	2.96	2.85
−0.8 - +0.4	7.45	91.46	6.24	1.22	0.15	8.69	2.65	4.54	3.34
−0.4 - +0.25	3.49	79.55	13.28	4.02	0.59	3.54	2.64	7.01	6.15
−0.25 - +0.12	7.01	74.19	16.99	3.81	0.62	6.64	6.79	13.35	12.98
−0.12 - +0.075	4.15	72.70	18.90	3.30	0.72	3.85	4.47	6.85	8.93
−0.075 - +0.043	4.81	73.17	19.41	2.91	0.76	4.49	5.33	7.00	10.92
−0.043	53.99	70.84	25.04	2.16	0.34	48.80	77.12	58.29	54.83
∑(original ores)	100	78.32	17.50	2.11	0.38	100	100	100	100

Table 4. Sorting results of the tailings by shaker.

Zone label	Yield /%	Gradation/%				Recovery ratio/%			
		SiO_2	Al_2O_3	Fe_2O_3	TiO_2	SiO_2	Al_2O_3	Fe_2O_3	TiO_2
Underflow K1	0.67	86.42	6.74	4.21	0.76	0.74	0.27	1.79	2.00
Middling K2	21.47	92.32	5.03	1.36	0.25	24.91	6.32	18.23	20.82
Middling K3	17.88	89.30	8.52	0.73	0.14	20.06	8.91	8.15	9.71
Middling K4	2.26	82.28	14.39	1.37	0.29	2.34	1.90	1.93	2.54
Overflow K5	57.71	71.64	24.46	1.94	0.29	51.95	82.60	69.90	64.92
∑(original ores)	100	78.32	17.50	2.11	0.38	100	100	100	100

Table 5. Sorting results of the tailings by settlement.

Product label	Gradation/%			
	SiO_2	Al_2O_3	Fe_2O_3	TiO_2
5 μm	59.71	31.64	1.38	0.14
10 μm	64.32	28.09	1.54	0.18
20 μm	65.61	24.47	2.5	0.41
Overflow in shaker	71.64	24.46	1.94	0.29

Table 6 shows the distribution of Fe element in the tailings. Fe element distributes mostly in titanomagnetite, kaolinite and limonite. The magnetic titanomagnetite and limonite can be removed by a magnetic separator. If it is done, the total Fe content in the tailing can be reduced to 1.23%, which meets the requirement of the standard of TC-3 in ceramic application. The effect of the magnetic induction on the Fe content is listed in Table 7.

As can be seen, the Fe gradation of the sorted tailing decreases with the increase of the magnetic field intensity. However, the ratio of iron removal is not proportional to the magnetic induction. The ratio of iron removal increases sharply with the magnetic field intensity. When the magnetic field intensity is 0.6T, the Fe gradation of the sorted tailing is 1.06% and the ratio of iron removal is 50.41%. Under this condition, most of titanomagnetite and limonite is removed. However, the ratio of iron removal increases slightly (4.68%) when the magnetic induction increases further from 0.6T to 1T. Therefore, the economical magnetic field intensity is 0.6T.

3.3. Separation Experiments in Pilot Scale

It has been proved that the tailing can be sorted by gravity method in laboratory scale. To make the tailing more profitable, kaolinite was separated from the tailing. The tailing slurry was firstly sieved by a high frequency vi brating sieve (GYX11-1210, sieve size is 0.4 mm). The coarse quartz was removed. The -0.44mm fraction of the slurry was delivered into a spiral chute (TGL 0610). Table 8 shows the separation results of the spiral chute.

Table 6. Distribution of Fe element in the tailings.

Mineral	Content of mineral/%	Fe_2O_3 content in mineral/%*	Fe_2O_3 distribution in tailing/%	Ratio of Fe_2O_3 distribution/%
Kaolinite	42.29	0.57	0.24	16.99
Quartz	49.69	0.17	0.08	5.98
Feldspar	3.90	0.16	0.01	0.43
Titanomagnetite	1.10	66.28	0.73	51.73
Limonite	0.29	53.19	0.15	10.85
Ilmenite	0.13	32.07	0.04	2.98
Muscovite	0.68	2.24	0.02	1.08
Biotite	1.36	10.34	0.14	9.98
Total	99.44	—	1.41	100

Notes: *the average value measured by MLA, which is smaller than by XRF due to the crystal water.

Table 7. Effect of the magnetic field intensity on the iron removal.

Magnetic field intensity/T	Fe_2O_3	
	Gradation/%	Ratio of iron removal/%
0.2	1.40	34.78
0.4	1.26	41.02
0.6	1.06	50.41
0.8	0.99	53.80
1	0.96	55.09
Tailing	2.14	—

As can be seen, the gradations of SiO_2 in Ks1 and Ks2 area are higher than the other areas, indicating that quartz can be sorted further by the spiral chute after removal of the coarse quartz. However, the fine quartz is also mixed with kaolinite because the recovery ratio of SiO_2 in Ks5 area is 43.40%. Correspondingly, kaolinite distributes mainly in Ks4 and Ks5 area, whose gradation is about 24% and total recovery ratio is 76.56%. The Al_2O_3 gradation still doesn't meet the standard of TC-3 due to the mixing of fine quartz.

To separate kaolinite from quartz further, the K4 and K5 fractions of the slurry were delivered into the hydrocyclone. The separation result is shown in Table 9.

It can be seen that the sorting by the hydrocyclone contributes to separation of kaolinite. However, the aim of the separation is still not obtained. The tailing in overflow needs to be separated second time. Table 10 gives the separation results of the hydrocyclone secondly.

Table 10 shows that the gradation of Al_2O_3 in the overflow is 28.6%. The sorted kaolinite can be applied in ceramic industries. Correspondingly, the yield of the overflow is 87.13%, implying that the separation by hydrocyclone secondly has the interesting in the point of economical view. The results in pilot scale agree with those in laboratory scale.

The whiteness is one of the most important factors in determining the application and economic value of kaolinite (Xia, et al. 2012) [12] . In general, the most deleterious impurities for whiteness of kaolin are iron oxide and titanium oxide minerals. Iron removal is required based on the results inTable 10 and the national standard of China (TC-3).

Iron removal of the tailing in overflow Kh2 is processed in high gradient magnetic separator (SSS-I-800). Table 11 gives the separation results of the magnetic separator.

Table 8. Separation results of spiral chute.

Production label	Yield/ %	Gradation/%				Recovery ratio/%			
		SiO_2	Al_2O_3	Fe_2O_3	TiO_2	SiO_2	Al_2O_3	Fe_2O_3	TiO_2
Underflow Ks1	16.78	89.13	7.54	3.13	0.61	19.19	5.91	23.81	26.06
Middling Ks2	10.12	86.57	10.52	1.00	0.27	11.25	5.74	4.59	6.96
Middling Ks3	12.03	78.76	18.19	1.19	0.27	12.17	11.79	6.49	8.27
Middling Ks4	15.07	72.29	24.00	2.50	0.40	13.99	19.49	17.09	15.35
Overflow Ks5	46.01	73.46	23.02	2.30	0.37	43.40	57.07	48.01	43.36

Table 9. Separation results of hydrocyclone (first time).

Product label	Yield /%	Gradation/%				Recovery ratio/%			
		SiO_2	Al_2O_3	Fe_2O_3	TiO_2	SiO_2	Al_2O_3	Fe_2O_3	TiO_2
Underflow Kh1	30.75	72.17	24.66	2.45	0.42	31.48	29.35	27.63	37.56
Overflow Kh2	69.25	70.93	26.36	2.85	0.31	68.52	70.65	72.37	62.44
∑(original tailing)	100	73.17	23.26	1.87	0.29	100	100	100	100

Table 10. Separation results of hydrocyclone (second time).

Product label	Yield/ %	Gradation/%				Recovery ratio/%			
		SiO_2	Al_2O_3	Fe_2O_3	TiO_2	SiO_2	Al_2O_3	Fe_2O_3	TiO_2
Underflow Kh1	12.87	72.46	24.95	2.66	0.33	13.66	11.41	11.82	13.59
Overflow Kh2	87.13	68.63	28.60	2.93	0.31	86.34	88.59	88.18	86.41
∑(original tailing)	100	70.93	26.36	2.85	0.31	100	100	100	100

Table 11. Separation results of the high gradient magnetic separator.

Product	Yield/ %	Gradation/%				Recovery ratio/%			
		SiO_2	Al_2O_3	Fe_2O_3	TiO_2	SiO_2	Al_2O_3	Fe_2O_3	TiO_2
Magnetic mineral	18.52	68.38	19.29	10.53	0.87	18.70	12.97	66.42	51.00
Diamagnetic mineral	81.48	67.57	29.43	1.21	0.19	81.30	87.03	33.58	49.00
Σ(original tailing)	100	68.63	28.60	2.93	0.31	100	100	100	100

The results in Table 11 show that 66.42% of Fe_2O_3 can be removed by the magnetic separator. Comparing to the results in Table 6, all the magnetic titanomagnetite and limonite have been removed by the magnetic separator. The magnetic separator in pilot scale has the same separation action with that in laboratory scale. The Fe_2O_3 content decreases from 2.93% to 1.21%, which is less that the requirement of TC-3 (1.8%). Obviously, the Fe_2O_3 that exists in diamagnetic minerals must be removed by chemical method if the quality of kaolinite needs to be increased further (Svoboda, and Fujita, 2003, Santos, et al. 2012, Chandrasekhar and Ramaswamy, 2007, Xia, et al. 2012, Zegeye, et al., 2013) [6] [10] - [13] .

In a word, after the above-mentioned separation processes, 30.2% of the ion-adsorption rare earth tailing can be re-sourced and applied in ceramic industries [The total yield is estimated by the yields in every separation process: (46.01 + 15.07)% × 69.25% × 87.13% × 81.84% = 30.2%]. The recovery ratio of kaolinite is 41.7%, which is estimated as the yield. Some of kaolinite cannot be sorted because it is mixed or accompanied by the other minerals. The recovery ratio of kaolinite may be increased by crushing the tailing. However, it is not economic considering most of the coarse grains in the tailing are quartz.

4. CONCLUSIONS

Kaolinite had been sorted from ion-adsorption rare earth tailings in Southern China in laboratory scale and pilot scale. The conclusions can be drawn as following:

1) The ion-adsorption rare earth tailing in Southern China is mainly compose of kaolinite and quartz, which can be re-sourced because no radioactive minerals exist.

2) Coarse quartz and fine kaolinite can be separated by sieves, shaker, spiral chute or hydrocyclone. There are the similar results in laboratory scale and in pilot scale. 30.2% of the tailings can be re-sourced and applied in ceramic industries.

3) 41.7% of kaolinite can be obtained after sorting and iron removal by magnetic separator in pilot scale, which meet the national standard of China (TC-3). Fe_2O_3 in the sorted kaolinite has the gradation of 1.21%, which should to be removed by chemical method if the quality of kaolinite needs to increase further.

ACKNOWLEDGEMENTS

The authors gratefully acknowledge the financial support provided by the National High-tech Research and Development Program of China (Grant No. 2012AA061903) and the National Natural Science Foundation of China (No. 51262013, 51362015).

REFERENCES

1. Su, W. (2009) Economic and Policy Analysis of China's Rare Earth Industry (in Chinese). China Financial and Economic Publishing House, Beijing.
2. Chen, Z. (2011) Global Rare Earth Resources and Scenarios of Future Rare Earth Industry. Journal of Rare Earths, 29, 1-6.
3. Yang, X. (2013) China's Ion-Adsorption Rare Earth Resources, Mining Consequences and Preservation. Environmental Development, 8, 131-136.
4. Wübbeke, J. (2013) Rare Earth Elements in China: Policies and Narratives of Reinventing an Industry. Resources Policy, 38, 384-394.

5. Chen, L. (2011) Effect of Magnetic Field Orientation on High Gradient Magnetic Separation Performance. Minerals Engineering, 24, 88-90.

6. Svoboda, J. and Fujita, T. (2003) Recent Developments in Magnetic Methods of Material Separation. Minerals Engineering, 16, 785-792.

7. Grimshaw, R.W. (1971) Physics and Chemistry of Clay. 4th Edition, Ernest Benn, London. [Paper reference:1]

8. Chandrasekhar, S. and Ramaswamy, S. (2006) Iron Minerals and Their Influence on the Optical Properties of Two Indian Kaolins. Applied Clay Science, 33, 269-277.

9. Malengreau, N., Bedidi, A., Muller, J.P. and Herbillions, A.J. (1996) Spectroscopic Control of Iron Oxide Dissolution in Two Ferralitic Soils. European Journal of Soil Science, 47, 13-20.

10. Santos, E., Scorzelli, R.B., Bertolino, L.C., Alves, O.C. and Munayco, P. (2012) Characterization of Kaolin from the Capim River Region—Brazil. Applied Clay Science, 55, 164-167.

11. Chandrasekhar, S. and Ramaswamy, S. (2007) Investigation on a Gray Kaolin from South East India. Applied Clay Science, 37, 32-46.

12. Xia, G., Lu M, Su, X. and Zhao, X. (2012) Iron Removal from Kaolin Using Thiourea Assisted by Ultrasonic Wave. Ultrasonics Sonochemistry, 19, 38-42.

13. Zegeye, A., Yahaya, S., Fialips, C.I., White, M.L., Gray, N.D. and Manning, D.A.C. (2013) Refinement of Industrial Kaolin by Microbial Removal of Iron-Bearing Impurities. Applied Clay Science, 86, 47-53.

9

CHAPTER

Dissolution of Rare Earth Elements from Coal Fly Ash Particles in a Dilute H2SO4 Solvent

Shunsuke Kashiwakura[1*], Yuichi Kumagai[2], Hiroshi Kubo[3], Kazuaki Wagatsuma[1]

[1]*Institute for Materials Research, Tohoku University, Sendai, Japan*

[2]*Soma Environmental Service, Co., Ltd., Soma, Fukushima, Japan*

[3]*Technical Research Institute of Obayashi Corpolation, Kiyose, Tokyo, Japan*

ABSTRACT

Recently, the worldwide supply of rare earth element (REE) resources will be severely restricted. On the other hand, coal fly ash particles emitted from coal-fired electric power plants contain relatively high concentrations of REEs. The contents of REEs in coal fly ash are regularly several hundreds of ppmw. In order to extract and recover REEs from coal fly ash particles, as a first step, we have investigated their dissolution behavior in a dilute H_2SO_4 solvent. The REE content of coal fly ash specimens has been precisely determined, and their presence in the ash component of the original coal and their enrichment in coal fly ash particles during coal combustion have been suggested. REEs in coal fly ash dissolve gradually in H_2SO_4 over time, and this implies two types of occurrences of the REEs in coal fly ash particles. By applying the unreacted core model to the dissolution behavior of REEs in a H_2SO_4 solvent, we can explain both types of occurrences.

Keywords: coal fly ash; rare earth elements; dissolution behavior; unreacted core model

1. INTRODUCTION

Currently, electricity generated from coal combustion accounts for around 25% of the total power generation in Japan. Since the energy supply, including electricity, is heavily dependent on imported resources, coal occupies an important position in Japan's electricity generation. Since a reduction in the proportion of electricity generated by nuclear power stations is unavoidable, owing to the disaster at the Fukushima Daiichi nuclear power stations, the utilization of coal-fired thermal electric power stations will become more important.

However, coal-fired power plants are recognized as an emission source of regulated substances such as coal ash; this is a hazardous industrial waste emitted from power plants, and in Japan, 10 Mt of coal ash is emitted per year [1]. Coal fly ash accounts for around 85% - 95% of the total coal ash emission [2] and has been widely recycled and used mainly as a raw material of fly ash cement in order to minimize landfill and to comply with the Basic Act for the Promotion of the Recycling-Oriented Society, enforced since 2000 [3].

Coal fly ash contains various trace elements [4], some of which are recognized to be hazardous. The hazardous inorganic substances include arsenic, boron, cadmium, fluorine, hexavalent chromium, mercury, lead, and selenium; their contents and the elution concentrations are strictly regulated by the Environmental Basic Act and its corresponding legislation [5,6].

Hence, the reduction of the hazardous components eluted from coal fly ash particles is necessary for their recycling. For this purpose, we have developed an acidwashing process [7-9]. The aim of this process is to remove the above-mentioned soluble regulated substances, focused on in our previous studies, from coal fly ash by dissolving them in an acid solvent. We showed that their dissolution, adsorption, and re-elution behavior depends strongly on the pH of the acid solutions and were successful in decreasing elution concentrations below regulatory limits.

In recent years, rare earth elements (REEs) have been recognized as important materials since they are widely utilized in the production of magnets, fluorescent materials, laser crystals, etc. [10]. Since Japan has been the largest consumer of REE products, exploration into alternative supply routes is important for resource security.

Coal fly ash contains relatively high concentrations of REEs. For example, NIST SRM 1633b and IRANT EOP, which are the standard reference materials for coal fly ash, contain 420 $mg \cdot kg^{-1}$ and 806 $mg \cdot kg^{-1}$ of REEs, respectively. Although these values are lower than those in ion-adsorbed type ore, they are equivalent to monaziteor bastnaesite-type ore [11].

REEs are known to be soluble in an acidic solvent. **Figure 1** shows the Pourbaix diagram for europium, as a typical example of a REE. It is clear that europium is soluble in acid as Eu^{3+}, implying that REEs may be recovered from coal fly ash particles. In this paper, we will discuss the dissolution behavior of REEs in dilute acidic solvents in order to examine the feasibility of REE recovery using an acid-washing process.

2. MATERIALS AND METHODS

2.1. Coal Fly Ash Specimens Utilized in This Study

Three types of coal fly ash specimens, Ash-A, Ash-B, and IRANT EOP, have been used in this work, and their compositions are listed in **Table 1**. Ash-A and Ash-B were provided by an electrical power company in Soma City, Fukushima Prefecture, Tohoku Region, Japan. As in the case of general coal fly ash [12], these specimens mainly consist of quartz (SiO_2), mullite ($3Al_2O_3 \cdot 2SiO_2$), and amorphous solids. The average diameter of the ash specimens was several tens of micrometers. IRANT EOP was obtained as a standard reference material from the Institute of Radioecology and Applied Nuclear Techni niques, Slovakia, and it has certified concentrations of lanthanum, cerium, and europium, and informative concentrations for scandium, samarium, terbium, ytterbium, lutetium (with expanded uncertainties), dysprosium, gadolinium, neodymium, praseodymium, and yttrium (without expanded uncertainties).

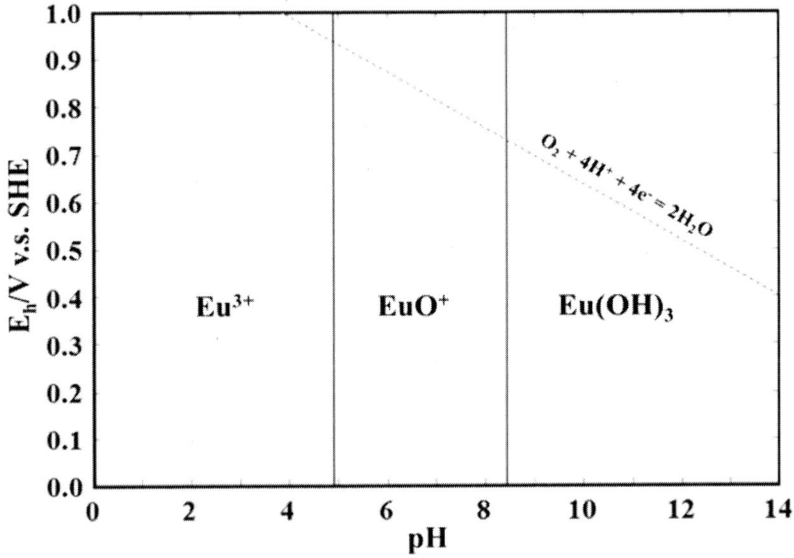

Figure 1. A pourvaix diagram for europium.

Table 1. Chemical compositions of coal fly ash specimens obtained by XRF analysis.

Contents /%	SiO_2	Al_2O_3	Fe_2O_3	CaO	MgO	Na_2O	K_2O	TiO_2
Ash-A	71.8	21.7	1.66	1.07	0.42	0.38s	1.36	0.75
Ash-B	62.2	20.8	3.44	8.52	1.57	0.37	1.14	0.65
Ash-EOP	49.1	29.8	7.37	2.35	0.97	0.50	0.77	6.02

2.2. Reagents and Experimental Procedure

In order to determine the REE concentrations in Ash-A and Ash-B and to verify the accuracy of the measurements obtained by inductively coupled plasma mass spectrometry (ICP-MS), the REE concentrations in the specimens were determined by means of acid digestion. Around 0.1 g of each specimen was put into a Teflon beaker. Reagent grade concentrated hydrochloric acid (35 wt%), nitric acid (60 wt%), and hydrofluoric acid (48 wt%), 10 ml each, were added. Two solutions were prepared for each ash specimen. They were then heated at 200°C for 6 h on a hot plate. After air-cooling, 10 mL of H_2SO_4 (95 wt%) were added and heated again to remove any fluorine, originating from the hydrofluoric acid, by vaporization of SiF_4(g). After confirming the generation of H_2SO_4fumes, indicating the complete vaporization of the fluorine content, they were again air-cooled. At this time, white residual dross, which may be $CaSO_4$, was observed in the solutions of Ash-B. The solutions were moved to 100 mL volumetric flasks, and at this point, the residual dross in the solutions had completely dissolved. After complete digestion, with the dissolution of the white residual dross visually confirmed, the solutions were prepared for measurement with inductively coupled plasma-mass spectrometry (ICP-MS; HP4500, Agilent) using an internal standard technique involving the addition of 0.2 $mg \cdot L^{-1}$ of rhenium.

The equipment for the dissolution test comprised a Pyrex beaker, a magnetic stirrer, a heater with a stirring magnet, electrodes that were used to monitor the pH, redox potential, and temperature of the solutions, and a laptop computer. Five grams of each coal fly ash specimen were agitated with 500 mL of the H_2SO_4 solvent, which was diluted at 10 times using ultrapure water. The solutions were agitated at 30°C, 60°C, and 80°C for 2 h. During agitation, the solutions were periodically sampled at 5, 10, 30, 60, and 120 min. The sampling of the solutions was conducted using a 10 mL macropipette. The aliquot was then filtered within 30 s, and 5 mL of each filtrate was diluted to 50 mL in a volumetric flask. ICPMS measurements were then conducted for the solutions with the rhenium internal standard described earlier.

3. RESULTS

3.1. The Determination of REE Elements in Coal Fly Ash

Table 2 summarizes the REE contents in IRANT EOP. The calibration line for the measured REEs was prepared in the range of 1 - 1000 μg L-1. The variance of all calibration lines is 1.00. The values of the REE concentrations are expressed as the mean value ± (2 × standard deviation). Since the true values of the REE concentrations are unknown, the recovery rate was calculated using averaged values as shown in Equation (1).

$$\text{Recovery rate} = \frac{\left(\text{Determined mean value in mg}\cdot\text{kg}^{-1} \times 100\right)}{\left(\text{Certified mean value in mg}\cdot\text{kg}^{-1}\right)} \tag{1}$$

The range of the calculated recovery rates are 92% - 118% for La, Ce, and Eu, for which IRANT EOP has certified values, and 72% - 133% for Sc, Sm, Tb, Yb, Lu, Y, Pr, Nd, Gd, and Dy, for which IRANT EOP has informative values. Iwata et al. [13] also determined REE components in the coal fly ash referenced standard material, NIST 1633b. Their recovery values were 48% - 123% for informative values, and 76% - 83% by comparison with the literature [14,15]. From this, the measured values of REE concentrations were considered to be reasonably well determined. The accuracy of the REE concentrations in coal fly ash determined using the acid-digestion method with hydrofluoric acid and subsequent ICP-MS measurement is considered to be adequate, and results for Ash-A and Ash-B are shown in **Table 3**.

3.2. The Dissolution Behavior of REEs in Coal Fly Ash Specimens

As typical examples, the dissolution behaviors of lanthanum, cerium, europium, and samarium in a dilute H_2SO_4 solvent are displayed in **Figure 2**. The dissolution rate was calculated by (dissolved mass of X in μg × 100/ content of X in μg), where the content of X was determined in the previous section. During the dissolution test, the values of the pH and redox potential for each solution were kept below 0 and above 800 mV (vs the standard hydrogen electrode (SHE)), respectively. The dissolution rates are seen to increase when the temperature of the H_2SO_4 solution increases, and the dissolution rates of Ash-B are higher than those of Ash-A. The other REEs, i.e., cerium, scandium, yttrium, praseodymium, neodymium, samarium, gadolinium, europium, terbium, dysprosium, holmium, erbium, thulium, ytterbium, and lutetium show dissolution behavior quite similar to lanthanum shown in **Figure 2**.

Table 2. Comparison of the determined REE contents in IRANT EOP.

	La	Ce	Eu		
EOP (This study, $mg \cdot kg^{-1}$)	148 ± 8	310 ± 3	5.1 ± 0.7		
EOP (Certified, $mg \cdot kg^{-1}$)	164 ± 9	322 ± 21	4.99 ± 0.77		
Recovery rate %	90	96	102		
	Sc	Sm	Tb	Yb	Lu
EOP (This study, $mg \cdot kg^{-1}$)	28.9 ± 2.3	17.4 ± 1.8	2.0 ± 0.38	2.7 ± 0.32	0.4 ± 0.06
EOP (Informative, $mg \cdot kg^{-1}$)	36.7 ± 3.6	21.9 ± 0.9	1.93 ± 0.59	4.41 ± 0.79	0.51 ± 0.04
Recovery rate %	79	79	103	62	76
	Y	Pr	Nd	Gd	Dy
EOP (This study, $mg \cdot kg^{-1}$)	34.3	29.4	112	21.1	8.3
EOP (Informative, $mg \cdot kg^{-1}$)	30	30	150	20	10
Recovery rate %	114	98	75	106	83

Table 3. Determined REE contents in Ash-A and Ash-B.

Contents $mg \cdot kg^{-1}$	Sc	Y	La	Ce
Ash-A	12.7	71	49.8	123
Ash-B	13.5	51	54.2	117
Contents $mg \cdot kg^{-1}$	Pr	Nd	Sm	Eu
Ash-A	14.5	62	16.8	3.2
Ash-B	13.4	54	10.8	2.4
Contents $mg \cdot kg^{-1}$	Gd	Tb	Dy	Ho
Ash-A	15.6	2.5	14.4	2.7
Ash-B	9.3	1.5	8.8	1.7
Contents $mg \cdot kg^{-1}$	Er	Tm	Yb	Lu
Ash-A	8.7	1.0	7.7	1.0
Ash-B	5.7	0.7	5.4	0.7

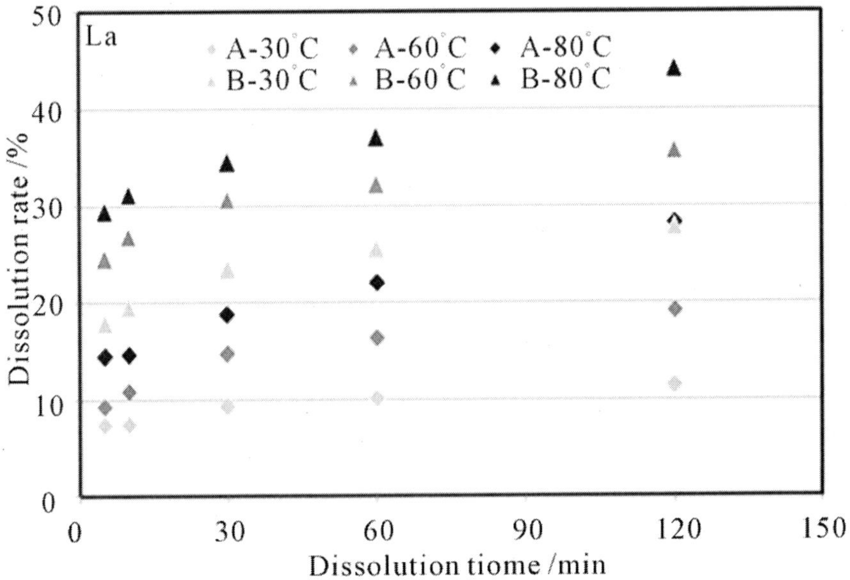

Figure 2. Changes in dissolution rates of lanthanum from Ash-A or Ash-B in a dilute H_2SO_4 solvent.

4. DISCUSSIONS

4.1. The Probability of Surface Deposition of REEs

The distribution of lanthanides in rocks or ores are known to follow the Oddo-Harkins rule; i.e., that atoms with even atomic numbers are more stable than those with odd numbers because their nuclei are more tightly bound. **Figure 3** shows the lanthanide contents in three coal fly ash specimens, and they completely conform to this rule since the lanthanides with even atomic numbers are always more abundant than neighboring lanthanides with odd atomic numbers.

In **Figure 4**, Masuda-Coryell plots [16,17] are shown for the REEs in the three coal fly ash specimens used in this study and the REEs in the original coal, for which the contents are taken from Fujita and Arikawa [18]. The relative abundances of the REEs were normalized with the REE contents in CI Chondrite [19]. In all plots, negative europium anomalies, derived from the progress of differentiation and found in various natural rocks, were clearly observed. The trends observed in the plots suggest that the REEs are distributed in the ash components of the original coal and then transferred into coal fly ash particles. Though we cannot directly compare the values of the relative abundances in coal and in the coal fly ash specimens plotted in **Figure 4**, the relative abundances of REEs in coal fly ash specimens are around 10 times those of REES in the original coals, suggesting that the coal fly ash particles are enriched with most of the REE content because after combustion, the weight of the coal fly ash is usually about one tenth of that of the original coal.

Two routes of the transition of REEs from the original coal to the coal fly ash particles can be considered. The first is widely known as a "vaporization surface-deposition process" [20,21]. Trace elements in the original coals, i.e., the semivolatile compounds [22-24], are vaporized from the original coal during combustion. After the formation of coal fly ash particles, the vaporized elements are deposited onto the surface of these particles.

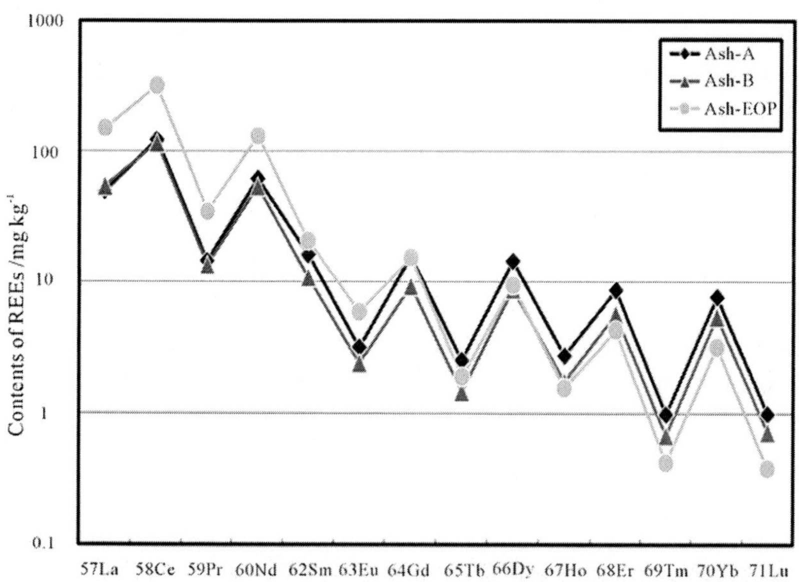

Figure 3. REE contents in coal fly ash specimens in this study.

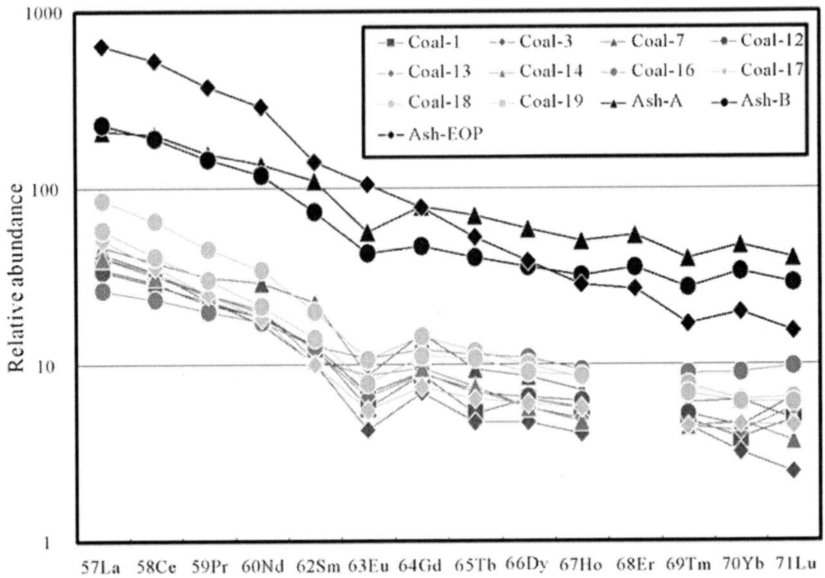

Figure 4. Masuda-Coryell plots for coal fly ash specimens and original coal from the reference.

In order to investigate the probability of vaporization of REEs, thermodynamically stable compounds of REEs were considered using a thermodynamic equilibrium calculation performed using HSC Chemistry 5.1 [25]. In this calculation, europium was selected as a typical REE. Forty-three species, with carbon, hydrogen, oxygen, nitrogen, sulfur, chlorine, fluorine, and europium contents set to their typical values for coal combustion, i.e., 60, 30, 60, 220, 0.1, 0.01, 0.002, and 0.00003 mol, respectively, were selected [26,27]. The ash components were ignored in order to focus on the vaporization behavior of europium and simplify the calculation. The total pressure was set to 1 bar, and the activity coefficient for each phase was assumed to be unity.

The result of the calculation is shown in **Figure 5**, from which it is confirmed that the volatility of $EuCl_3$(g) is much higher than its fluoride, oxide, or elemental vapor. The vaporized $EuCl_3$(g) may then be deposited on the surface of coal fly ash particles.

4.2. Different Types of Occurrences of REEs in Coal Fly Ash Particles

The other type of REE transition is their accompaniment of blown-up ash components from the original coal. In this case, REEs were melted with ash components and distributed into the coal fly ash particles, such as quartz (SiO_2), mullite ($3Al_2O_3{\cdot}2SiO_2$), and an amorphous phase. Since coal fly ash particles appeared to be spherical and consisted of a glassy phase together with crystalline phases, such as quartz and mullite, we have assumed both types of occurrences of REEs, namely surfacedeposition and accompaniment. From this assump tion, the dissolution behavior of REEs in a dilute H_2SO_4 solvent was fitted using the unreacted core model [28]. Coal fly ash particles were assumed to be spherical particles with uniform sizes. If we assume the occurrence of a diffusion process in the reacted layer or a chemical reaction at the surface of the unreacted core as a rate-determining step, then the following well-known rate equations are obtained.

$$\left(\frac{dn}{dt}\right)_D = 4\pi r D \frac{dC_a}{dr}$$ (interlayer diffusion control) (2)

$$\left(\frac{dn_i}{dt}\right)_C = 4\pi r^2 k_c C_a$$ (interfacial reaction control) (3)

Here, n_i, t, D, C_a, r, and k_c denote the number of moles of unreacted element i (mol), leaching time (s), diffusion coefficient ($m_2{\cdot}s^{-1}$), acid concentration ($mol{\cdot}L^{-1}$), radius of the unreacted core (m), and interfacial reaction rate constant ($m{\cdot}s^{-1}$), respectively. By integrating and rearranging Equations (2) and (3), the following well-known rate equations are obtained.

$$3\{1-(1-f)^{2/3}\}-2f = at$$ (interlayer diffusion control)(4)

$$1-(1-f)^{1/3} = bt$$ (interfacial reaction control) (5)

Here, a and b are defined as the inverse of the termination times of the reactions, and we can treat them as proportionality constants. **Figure 6** shows the fitting results for the dissolution of lanthanum as a typical example. When we assume their dissolution is controlled by interlayer diffusion, the unreacted core model can fit the dissolution rate very well with a variance greater than 0.87. The intercepts of all of the fitting lines are non-zero. Similar trends are found in the case of the rest of the REEs considered in this study. Therefore, these fitting lines suggest two types of occurrences of REEs, i.e., in the early stages of dissolution, REEs at the surface of the coal fly ash dissolve immediately, resulting in the drastic increase in the $(3(1 - f)^{2/3} - 2f)$ value. Thereafter, REEs inside the coal fly ash particles dissolve gradually, corresponding to the linearity of the fitting line.

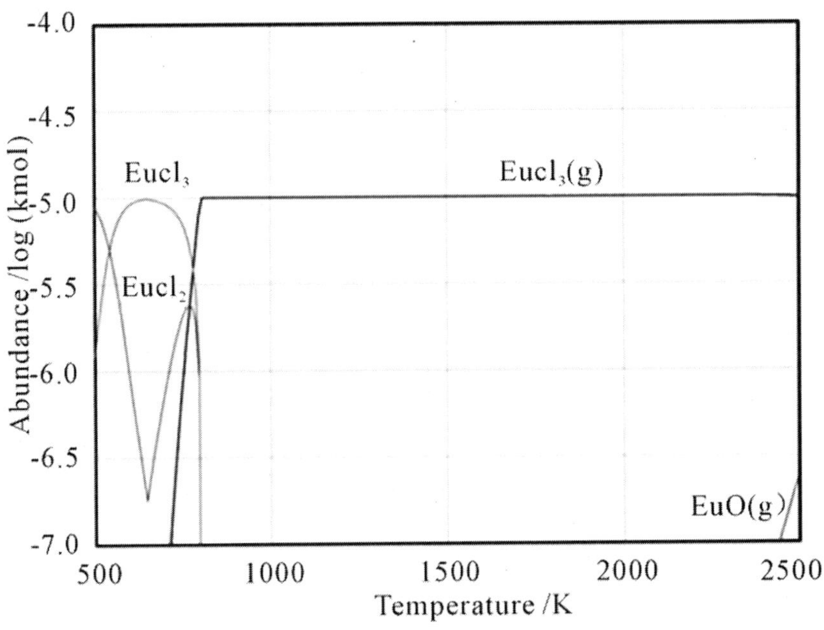

Figure 5. The result of thermodynamic equilibrium calculations for several europium species in coal.

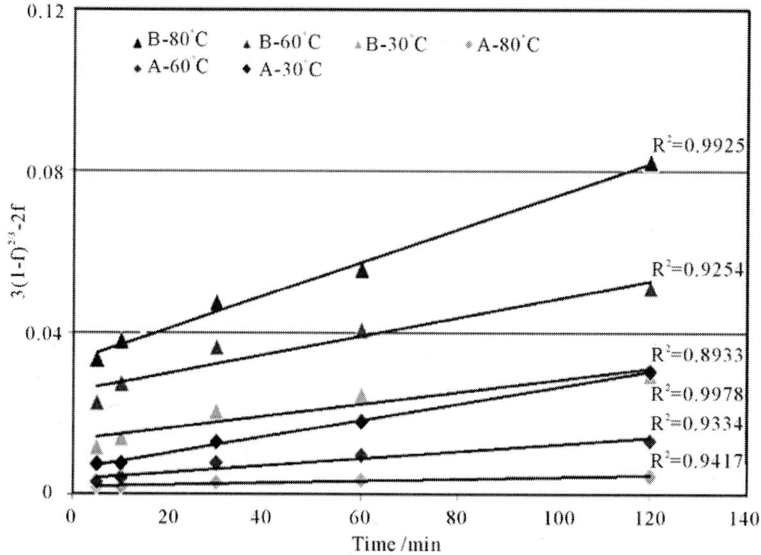

Figure 6. Dissolution rates of lanthanum in a dilute H_2SO_4 solvent fitted by the unreacted-core model assuming that dissolution is controlled by interlayer diffusion.

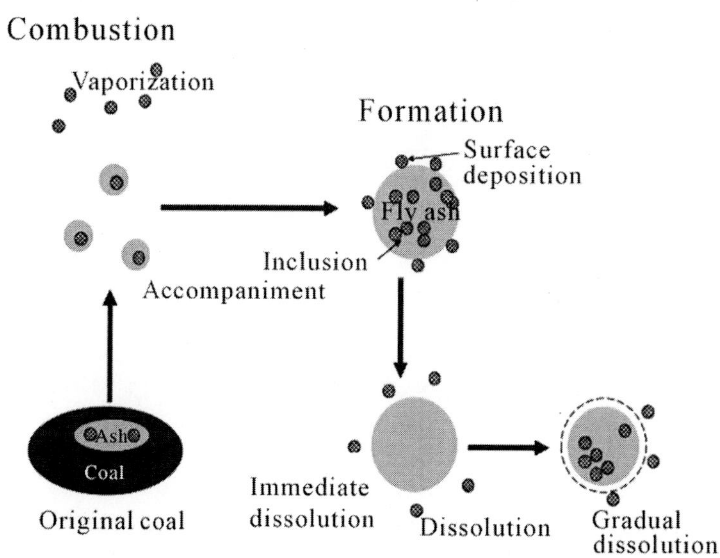

Figure 7. The estimated pass of REEs.

The estimated transition pass of REEs through coal combustion and dissolution in H_2SO_4 were described in **Figure 7**, where the REEs were drawn as checked circle The degree of "surface-deposition" and "accompaniment" will be greatly affected by the type of the original coal and combustion conditions, including temperature, oxygen concentration, residence time, and cooling rate of coal fly ash particles, and their effects on the degree of each type of occurrences of REEs will be investigated in future work.

5. CONCLUSION

A Laboratory-scale dissolution experiments were conducted to determine REE contents in coal fly ash particles and investigate their dissolution behavior in H_2SO_4 solvents. The REE contents were precisely determined by comparing the results with IRANT EOP, which were considered as the benchmark for accuracy. In the dissolution tests, dissolution rates of REEs increased with the temperature of the H_2SO_4 solvent. The REE content in coal fly ash specimens completely conformed to OddoHarkins rule, and the Masuda-Coryel plots show that most of them are probably derived from the ash component of the original coal. The transition of the REEs was attributed to surface deposition and accompaniment, and the probabilities of both processes could be explained by applying the unreacted core model to the dissolution behavior of REEs from coal fly ash in H_2SO_4solvent.

6. ACKNOWLEDGEMENTS

The authors would like to express their sincere gratitude to the Grant-in-Aid for Scientific Research (KAKENHI) 23760710 from the Ministry of Education, Culture, Sports, Science and Technology, Japan, and to the Steel Foundation for Environmental Protection Technology for their financial support. We also acknowledge the Technical Division, School of Engineering, Tohoku University, and the Analytical Research Core for Advanced Materials, Institute for Materials Research, Tohoku University, for facilitating the ICP-MS measurements.

REFERENCES

1. Japan Coal Energy Center, "Coal Ash Database".
2. Societies of Civil Engineers, "The Technologies about Effective Utilization for Coal Fly Ash Aiming for the Sound Material-Cycle Society," Japan, 2003.
3. Ministry of the Environment, "Environmental Basic Act," Japan, 1993.
4. M. Xu, R. Yan, C. Zheng, Y. Qiao, J. Han and C. Sheng, "Status of Trace Element Emission in a Coal Combustion Process: A Review," Fuel Processing Technology, Vol. 85, No. 2-3, 2003, pp. 215-237.
5. Ministry of the Environment, "Fundamental Law for Establishing a Sound Material-Cycle Society," Japan, 2000.
6. Ministry of the Environment, "Soil Contamination Countermeasures Act," Japan, 2002.
7. S. Kashiwakura, H. Kubo, Y. Kumagai, H. Kubo, K. Matsubae-Yokoyama, K. Nakajima and T. Nagasaka, "Removal of Boron from Coal Fly Ash by Washing with HCl Solution", Fuel, Vol. 88, No. 7, 2009, pp. 1245-1250.
8. S. Kashiwakura, H. Ohno, K. Matsubae-Yokoyama, Y. Kumagai, H. Kubo and T. Nagasaka, "Removal of Arsenic in Coal Fly Ash by Acid Washing Process Using Dilute H_2SO_4 Solvent," Journal of Hazardous Materials, Vol. 181, No. 1-3, 2010, pp. 419-425.
9. S. Kashiwakura, H. Ohno, Y. Kumagai, H. Kubo, K. Matsubae and T. Nagasaka, "Dissolution Behavior of Selenium from Coal Fly Ash Particles for the Development of an Acid-Washing Process," Chemosphere, Vol. 85, No. 4, 2011, pp. 598-602.
10. JOGMEC, "Material Flow of Mineral Resources," 2010.
11. Z. X. Yuan, C. Y. Wu, L. M. Xu and Y. X. Ni, "The Distribution of Trace Elements in Granitoids in the Nanling Region of China," Chinese Journal of Geochemistry, Vol. 12, No. 3, 1993, pp. 193-205.

12. S. Kashiwakura, T. Takahashi, H. Maekawa and T. Nagasaka, "Application of 11B MAS-NMR to the Characterization of Boron in Coal Fly Ash Generated from Nantun Coal," Fuel, Vol. 89, No. 5, 2010, pp. 1006-1011.

13. S. Iwata, K. Minamoto, E. Fujimori, K. Chiba and H. Haraguchi, "Determination of Rare Earth Elements in Industrial Waste Incineration Fly Ash by ICP-MS and the Characteristics of the Rare-Earth Element Distribution Pattern (in Japanese)," BUNSEKI KAGAKU, Vol. 50, No. 6, 2001, pp. 419-425.

14. R. R. Greenberg, E. A. Mackey and D. A. Becker, "The Application of Instrumental Neutron Activation Analysis for the Certification of the New NIST Fly Ash SRM," Journal of Radioanalytical and Nuclear Chemistry, Vol. 193, No. 1, 1995, pp. 7-14.

15. I. Roelandts and E. S. Gladney, "Consensus Values for NIST Biological and Environmental Standard Reference Materials," Fresenius Journal of Analytical Chemistry, Vol. 360, No. 3-4, 1998, pp. 327-338.

16. A. Masuda, "Regularities in Variation of Relative Abundances of Lanthanide Elements and an Attempt to Analyse Separation-Index Patterns of Some Minerals," The Journal of Earth Sciences, Nagoya University, Vol. 10, No. 2, 1962, pp. 173-187.

17. C. D. Coryell, J. W. Chase and J. W. Winchester, "A Procedure for Geochemical Interpretation of Terrestrial Rare-Earth Abundances Patterns," Journal of Geophysical Research, Vol. 68, No. 2, 1963, pp. 559-566.

18. K. Fujita and Y. Arikawa, "Determination of Rare Earth Elements in Coal Samples by ICP-MS," The Bulletin of Japan Women's University, Vol. 14, 2006, pp. 7-11.

19. W. McDonough and S. Sun, "The Composition of the Earth," Chemical Geology, Vol. 130, No. 3-4, 1995, pp. 223-253.

20. R. L. Davison, D. F. S. Natusch, J. R. Wallace and C. Evans Jr, "Trace Elements in Fly Ash. Dependence of Concentration on Particle Size," Environmental Science & Technology., Vol. 8, No. 13, 1974, pp. 1107-1113.

21. K. C. Galbreath, D. L. Toman, C. J. Zygarlicke and J. H. Pavlish, "Trace Element Partitioning and Transformations during Combustion of Bituminous and Subbituminous U. S. Coals in a 7-kW Combustion System," Energy & Fuel, Vol. 14, No. 6, 2000, pp. 1265-1279.

22. R. Meij, "Trace Element Behavior in Coal-Fired Power Plants," Fuel Processing Technology, Vol. 39, No. 1-3, 1994, pp. 199-217.

23. R. Meij and H. Winkel, "The Emissions of Heavy Metals and Persistent Organic Pollutants from Modern CoalFired Power Stations," Atmospheric Environment, Vol. 41, No. 40, 2007, pp. 9262-9272.

24. S. Kashiwakura, T. Takahashi and T. Nagasaka, "Vaporization Behavior of Boron from Standard Coals in the Early Stage of Combustion," Fuel, Vol. 90, No. 4, 2011, pp. 1408-1415.

25. A. Roine, "Outokumpu HSC Chemistry for Windows, Version 5.1," 2002.

26. N. Tsubouchi, Y. Wang and Y. Ohtsuka, "Fate of the Chlorine in Coal during Temperature-Programmed Pyrolysis," The proceedings of the Japan Institute of Energy, Vol. 40, 2003, pp. 124-125 (in Japanese).

27. N. Tsubouchi, H. Hayashi and Y. Ohtsuka, "Functional Forms of the Fluorine and Carbon in Fly Ashes Formed in Pulverized Coal Combustion," The proceedings of the Japan Institute of Energy, Vol. 42, 2005, pp. 59-60 (in Japanese).

28. S. Yagi and D. Kunii, "Fluidized-Solids Reactors with Continuous Solid Feed—I: Residence Time of Particles in Fluidized Beds," Chemical Engineering Science, Vol. 16, No. 3-4, 1961, pp. 364-371.

10

CHAPTER

Rare Metal Deposits of East Kazakhstan: Geologic Position and Prognostic Criteria

Boris Dyachkov[1,2], Natalya Zimanovskaya[1], Indira Mataibayeva[1]

[1]D. Serikbayev East Kazakhstan State Technical University, Ust-Kamenogorsk City, Kazakhstan

[2]Altai Geological and Ecological Institute, Ust-Kamenogorsk City, Kazakhstan

ABSTRACT

In the article the features of the formation and metallogeny of the geological structures of Great Altai (Rudny Altai, Kalba-Narym, Western Kalba and Jarma-Saur) which are included into the system of the Central Asian mobile belt are considered. The characteristic of the main types of rare metal minefields of the Kalba-Narymsky belt genetically connected with the Perm granitoid magmatism of the post-conflict orogenny stage of activization is given. The rhythmical and pulsation model of pegmatitovy ore formation in the conditions of the half-closed magmatic system, reflecting the phasic development of mineral complexes from graphic and oligoclase-microcline (barren) to microcline-albite and albite-spodumene ore (Ta, Nb, Be, Li, Sn, etc.) is developed. On the basis of the revealed criteria of ore formation recommendations about the direction of the further researches are made.

Keywords: Rare Metals; Pegmatite; Albitite-Greisen Metasomatits; Kalba-Narym Belt; East Kazakhstan

1. INTRODUCTION

The examined territory is a part of the Central-Asian mobile belt zone and unites geological structures of Great Altai including Rudny Altai, Kalba-Narym, Western Kalba both Jarma-Saur and abuted areas of Russia and China (**Figure 1**). Borders are the northwest deep breaks separating hercine structures of Great Altai from caledonide formations of Gorne Altai (in the northeast) and Chingiz-Tarbagatay (in the southwest). General extent of territory is more than 1000 km while 300 - 400 km at width.

The Kalba-Narym zone is composed of Upper Devonian and Carboniferous terrigenous rock that are intruded by the major Kalba-Narym granitic pluton. The main litologies are granodiorite, granite and leucogranite, aplitelike granite, aolite dikes, pegmatite and quartz veins. Early granitic rocks are assigned to the Kalba Complex (P_1), whereas the later leucogranite is referred to the Monastyrsky Complex (P_2).

Rare metal pegmatite deposits in the Central Kalba ore district (Yubileinoe, Belogorskoe, Verkhne-Baimurzinskoe, Ognevskoe, etc.) are of the most economic importance. Two deposits of the pegmatittype (Yubileinoe) and greisens-quartz vein type (Cherdoyak) are discussed in the text below.

Last years from new theoretical positions—the hypotheses of global mobilization problematic questions of geodynamic development of Great Altai main ore-bearing structures were considered. The results of researches are published in the monography "Great Altai" [1,2].

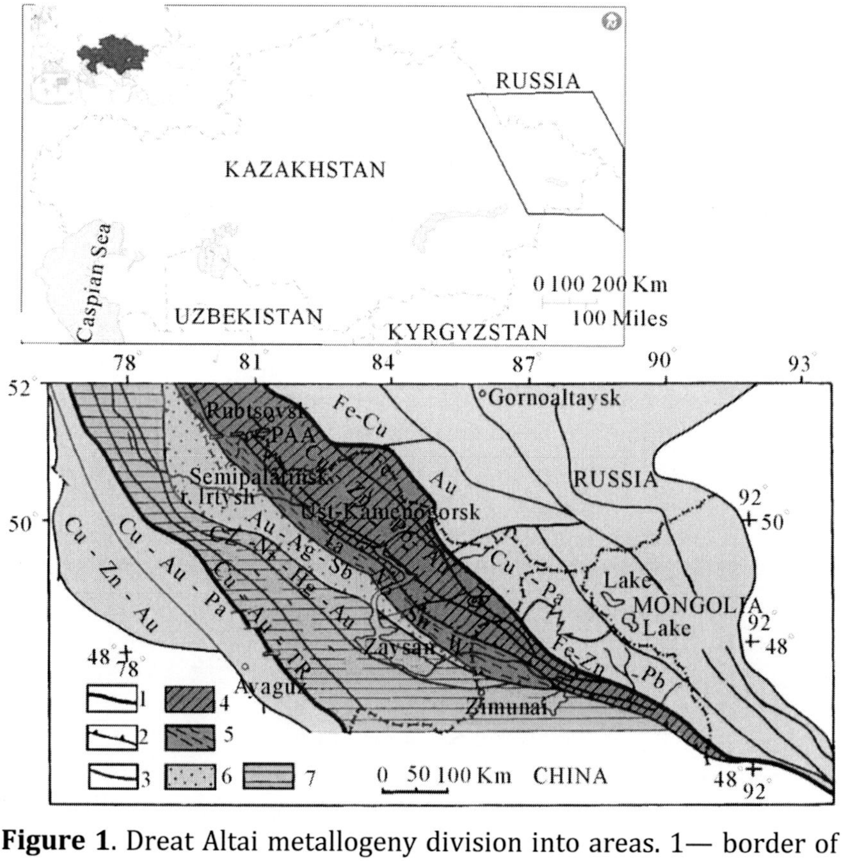

Figure 1. Dreat Altai metallogeny division into areas. 1— border of Dreat Altai; 2—of ore belts and 3—of metallogenic zones; 4—Rudnoaltaisky, 5—Kalba-Narymsky, 6— West-Kalbinsky and 7—Zharma-Saursky ore belts.

2. THE GEOLOGICAL AND METALLOGENIC EVOLUTION

The formation of geological structures contacts to general geodynamic model of Paleoasiatic ocean evolution and was defined by system of complex cooperating continental lages of the Siberian and Kazakhstani subcontinents. Modern geological structures of region are tectonic enclave of ancient paleocontinents and litospheric blocks (or collage terrains), which drifted in Paleasiatic ocean and have confronted in the process of hercinic collision (C_1-C_{2-3}), having formed uniform structure of Great Altai [3-5].

According to the region's metallogeny we allocate here four ore belts of a belt zone: Rudnoaltai (copperpolimetallic), Kalba-Narym (rare metals), West-Kalba (gold ore) and Zharma-Saur (multimetal) [2,6]. The geological formation development and mineralization occured in a mode of oceanic riftogene (PR_3), then in early (riftogenic, islandarcic), average (collision) and late (postcollision) stages of coledanic and hercynic cycles, and was finished by continental riftogenes and neotectonic orogeny in mesozoic and cainozoic [1,6]. In Precambrian during the process of proterozoic base riftogenes in mantled breaks magmatics deposits of a Cr, Ni, Co (Pt, Hg) were generated. In Rudny Altai in riftogenic stage (D_{1-3}) were formed large pyritespolimetallic deposits of volconogenic genesis—Cu, Pb, Zn, Au, Ag etc. (Zyryanovsk, Maleevsky, Ridder-Sokolny, Artemevsky etc.) [2,6-8]. For an average conflict stage (C_{1-3}) sulphidic copper-nickel, copper-porphiry and gold ores deposits connected with magmatigm of gabbro-dioritgranodiorit-plagiogranit series are characteristic. The most significant deposits of gold (Bakyrchik, Suzdal etc.) were located in Zaisan suturn zone generated on continental edges interfase [6,9,10]. In a late stage (P_1-T_1) under the conditions of innerslabs activation were generated granitoid belts (zone) with rare metals and rare earths ores, placed on lithospheric blocks borders. In cimmerian cycle in conditions of continental riftogenesis and subsequent stabilization the deposits of crustores (Ni-Co, Zr-Ti-Au) were formed.

Thus, main rare metal-bearing structures of granitoid region belt have appeared during late hercynion activization differing on geotectonic position, the scale of development, internal structure, material structure and orebearing (**Figure 2**). Natural spatial time of rare metals belts to tectonic weakened zones in continental blocks terrestrial crust of raised sialic is established. There is marked northwest line and significant extent (500 - 800 km).

3. KALBA-NARYM BELT

Kalba-Narym belt is the largest adjoining to Irtysh shear zone and extends in the northwest direction on 500 km and further proceeds in Russia and China. On the geology-geophysical data it is supposed, that this granitoid belt is placed in a head part of huge techtono-magmatic zone, steep fallic on the northeast under Rudny Altai [2]. The centers of magmaformation arose, by granite melts structure in metagranited layer or on its border with metadiorited layer. The transit ore-bearing solution have penetrated from the bottom parts of Earth crust and top mantle on system of deep breaks [2,11].

The belt is combined with mainly normal granits and leucogranits of the Perm age, to which the deposits of rare metals are generically connected (Ta, Nb, Be, Li, Cs, Sn, W) [12,13].

We allocate the following ore formations types of deposits:

1) pegmatitic rare metals (Ta, Nb, Be, Li, Cs, Sn), submitted by the main industrial deposits (Bakenny, Belaya Gora, Yubileiny etc.);
2) pegmatic beryl-microcline in which mineral deposits are blocked microcline and quartz, muscovite, beryl and columbite (deposits: Asubulak, Lobaksai, NijneLaibulak etc.);
3) albite-greisen tin—tantalitic (apogranitic) in the latent granite domes, potentially perspective on revealing Ta, Sn, Be, Li (Karasu deposit);
4) greisen-guartz vein cassiterite wolframic submitted by ores of wolframite, scheelite and cassiterite (Cherdoyak, Palasy, Kaindy deposits, etc.);
5) clastogene tantal-cassiterite-wolframic, forming tantalite looses, cassiterite, wolframite, scheelite and monazite.

3.1. The Model of Rare Metal Pegmatite Deposits

The basic model of formation for main industrial rare metal—pegmatite of Kalba-Narym belt deposits is defined by their genetic connection with granite kalba of complex P_1 and spatial accommodation of ores veins mainly in granite files and their exocontact (**Figure 3**). The processes of pegmatite formation probably occured in open or semi-closed magmatic system at rhythmic— pulsating receipt of ore-bearing distillates (H_2O, F, B, Cl, Ta, Nb, Be etc.) from cameric center of granite files (under the conditions of the increased activity of ore-controlling breaks). Pulsating receipt of ore-bearing solutions has determined multirhythmical zones of ash value pegmatic veins, the colour quartz—lepidolite-cleavelandite and spodumene with rich complex ores (Ta, Nb, Li, Cs, Be, Sn) and increasing concentration of mineralization veins stagic development of mineral complexes from oligoclase-microcline up to albite.

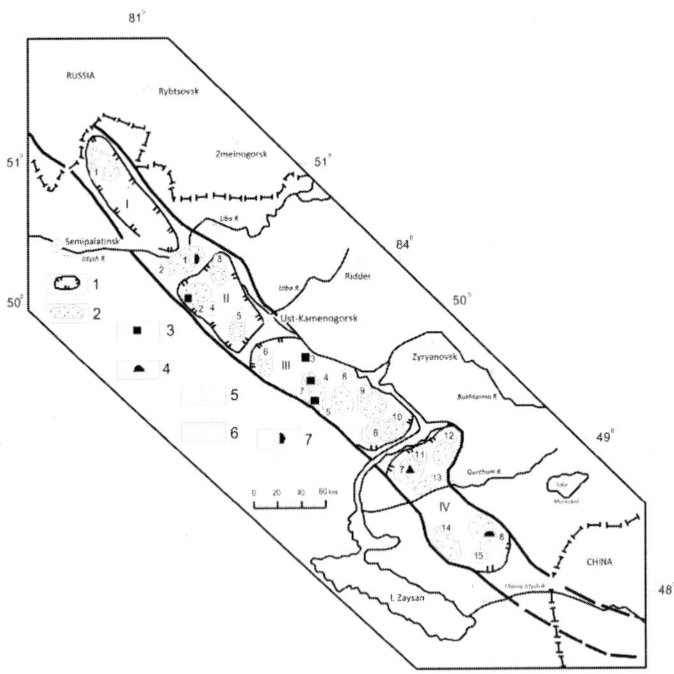

Figure 2. Kalba-Narym rare-metal belt [6]. Boundaries of (1) ore belts and (2) ore clusters; (3 - 7) types of deposits: (3) rare-metal pegmatites, (4) Snand Ta-bearing albitite and greisen, (5) greisens and quartz veins, (6) Sn-W quartz veins, (7) copper-pyrrhotite massive sulfide deposits. Ore districts (numerals in figure): I—Shulba; II—Northwest Kalba; III—Central Kalba; IV Narym. Ore clusters (numerals in circles): 1—Stepanovsky; 2—Vavilonsky (copper); 3-Mokhnatukhinsky; 4—Ubinsky; 5—Yuzhnoproletarsky; 6—Saryozek; 7—Asubulak-Belogorsky; 8—Manat-Chernovinsky; 9—Voilochevsky-Kosheminsky; 10—Mirolyubovsky; 11—Leninsky-Cherdoyak; 12—Kasatkinsky-Bulandinsky; 13—Burabai; 14—Tortqalmak; 15—Karasu; deposits (numerals in figure): 1—Vavilon; 2—Kvartsevy; 3—Bakenny;4—Yubileiny; 5—Belaya Gora; 6— Palatsy;7—Cherdoyak; 8—Karasu.

The carried out researches have shown, that ore-generating ability of granitoide, alongside with petrologic by the factors, in many respects depends on geodynamic conditions stand of files and scales degasazion ore-bearing solution melts [2,6]. From these positions certain oremagmatic systems differing on the scale miner were planned.

The above-intrusive system of active interaction granite intrusion with lateral breeds raised carbonaceous and calcindition in mobile tectonic conditions. Rare metal mineralization (Ta, Nb, Be and others) concentrated in a frontal part of stratified intrusivic bodies, in their apofise and branches, and also in lateral breeds of eksocontakte. The main industrial type is Bakenny deposit (**Figure 4**).

The inside-intrusic system of cameric conter formation of ore pegmatite veins in granite files complicated by explosive tectonic. Ore-controlling was the regional breaks of sublatitudic stream line, opening residual cameric of ore-bearing melts, and ore-locading were zones raising cracks and zones of increased fissing and independent system of cracks in granite at oreformation rare metal pegmatites of albite subformation, uniting various mineral complexes (albite-microcline, albite, albite-spodumene etc) were generated. Ore bodies contain the unique minerals clevelandit, coloured and polychrome tourmalines, lepidolite, spodumene, petalite, pollucite and othres. The deposits have industrial values (Yubileiny, Belaya Gora, Upper Baymurza etc.) [2,6,7, 14]. These objects are compared with the large pegmatite deposits in other regions-Bernic Lake (Canada), Koktokay (China) and other [15-17].

3.2. Types of Deposits

System of ore-bearing granite domes. It is characterized by concentration of Sn-Ta mineralization in albitic and greisenic granite intrusive-above-intrusive zone of a blind dome (type Karasu) (**Figure 5**). Rare metal mineralization is generically connected with medium musco vite phase of granites the II phase of Kalba complex. The latter on geochemical features are referred to Sn-W-Ta type of ore-bearing granite close to liti-ftoric granite according to W. I. Kovalenko (1977).

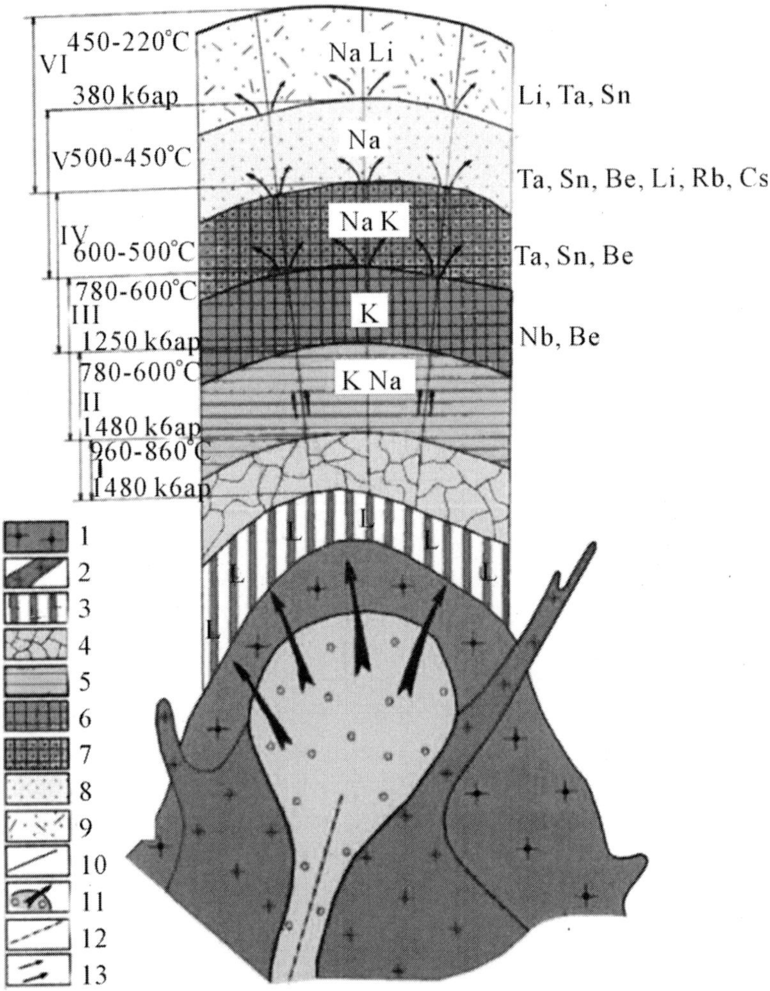

Figure 3. A model formation pegmatite rare deposits. 1— coarse- and medium-biotite granite; I phase and 2—fine-and medium-grained muscovitized granite, II phase kalba complex (P_1); 3—aplite-like, aplite; 4 - 10—pegmatites (4— oligoclase-microcline barren, 5—grafic, 6—microcline in blocks with beryl and columbite 7—microcline-albite with tantalite, cassiterite, beryl (rarely spodumene), 8—albitic (productive) with tantalite, cassiterite, beryl, pollucite, spodumene, lepidolite; 9—albite (cleavelandite)-spodumene, with tantalite, amblygonite, petalite, polychromatic tourmaline; 10—faults; 11, 13—direction of the fluid movement; 12—ore-controling deep fault.

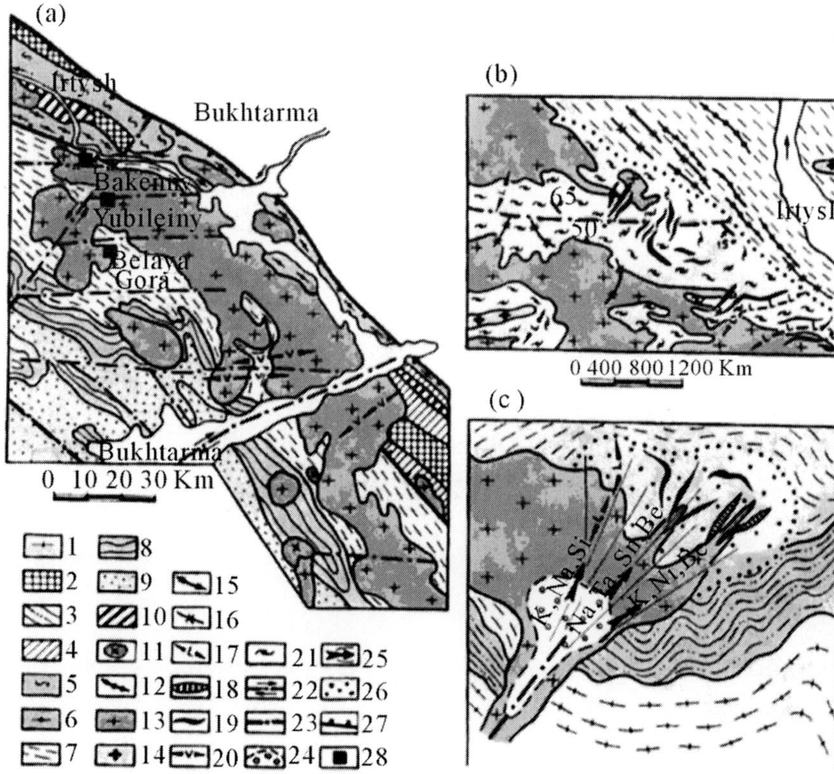

Figure 4. Bakennoe pegmatite deposit. (a)—arrangement plan of the pegmatite deposites; (b)—deological composion and (c)—geology-genetic model of the Bakennoe deposit. 1—granitic-metamorphic layer; 2—blocks of pregercynic basement; 3—calcic-terrigenous sediments of kystav-kurchumskoy series (D_2gv); 4 - 6—metamorfoseds sediments (D_{1-2}); 7—carbonaceous-sandstone-aleurolite sediments (D_3- C_1); 8—flyschoid formation (C_1v_{2-3}) and 9—grey-wacke; 10—gabbro-plagiogranite and migmatite-gneiss-granite formation (C_1); 11—plagiogranite-dranodiorites and 12— dikes of the kunushe complex (C_3); 13 - 19—kalba complex (13—granite I phase and 14—II phase, 15—aplite-like granite, 16—aplite-pegmatite, 17—oligoclase-microcline, 18 —microcline blocks and 19—rare metal pegmatite); 20 —gabbro-diabase-granite-porhyry formation (P_2-T_1); 21 —granitization rocks, migmatite; 22—fractures; 23—orecontrolling deep fault; 24—focus of ore bringer fusion; 25— direction of ore-bearing fluid travel; 26—emanation aureal of rare elements; 27—boundary of Irtysh crush zone; 28— pegmatits deposites (Using J. A. Sadovsky's materials).

The prognosic resources of tantalite and tin are significant.

System of imposed rare metal mineralization. It is characterized by migration of ore-bearing fluidizations from crystallizing granite melts in the tectonic weakened zones in containing breeds and imposing here mineralization on earlier hypabyssal small intrusions and dykes of Cunush complex C_3. The latter became favorable physics-chemical environment for adjournment and concentration of mineralization, acting in a role of structural —litologic and geochemical traps (deposits Medvedka, Tochka etc.).

The system field of dispersion RE is typical for large files of leucogranite Monastyr complex (P_2), which differ by flying and mineralizer riches (H_2O, F, B etc.) and geochemical specialization on Ta, Li, W, Sn, TR etc. However owing to crystalle-chemical dispersion of rare elements leucogranitic formation has appeared to be loose ore bearing. It is connected with chamber crustbearing pegmatite and fine greisene-quartzvein wolframic demonstration (deposit Dungaly).

Concerning rating of territory prospects it is possible to note, that Kalba-Narym belt is the main rare metal structure of Great Altai. Perspective are the northwest and southeast flanks granitoide of a belt where the detection of new rare metal deposits are latent on depth and hide under a cover of friable deposits. Further we'll briefly consider the characteristics of others granitoidic belts [6]. On the border of Great Altai with caledonion of Gorny Altai Tigirekskо-Chernevinsky belt of normal and increased alkali granitoide was generated. The main here are skarnic and greizic deposits with W-Mo (Ekipetsky, Ivanovsky etc.), and in alkaline granite Nb-Zr—rare earths mineralization is shown (Azutau). In Gorny Altai and Chinese Altai in zones of hercinic tectone-magmatic activization Uryl-Koktokay belt of normal and subalkaline granite with pegmatite Ta-Nb-Be-Li is allocated with greisene and hydrothermal W-Mo-Be deposits (Koktokay, Kalguty, Kokkol etc.) and displays rare earths

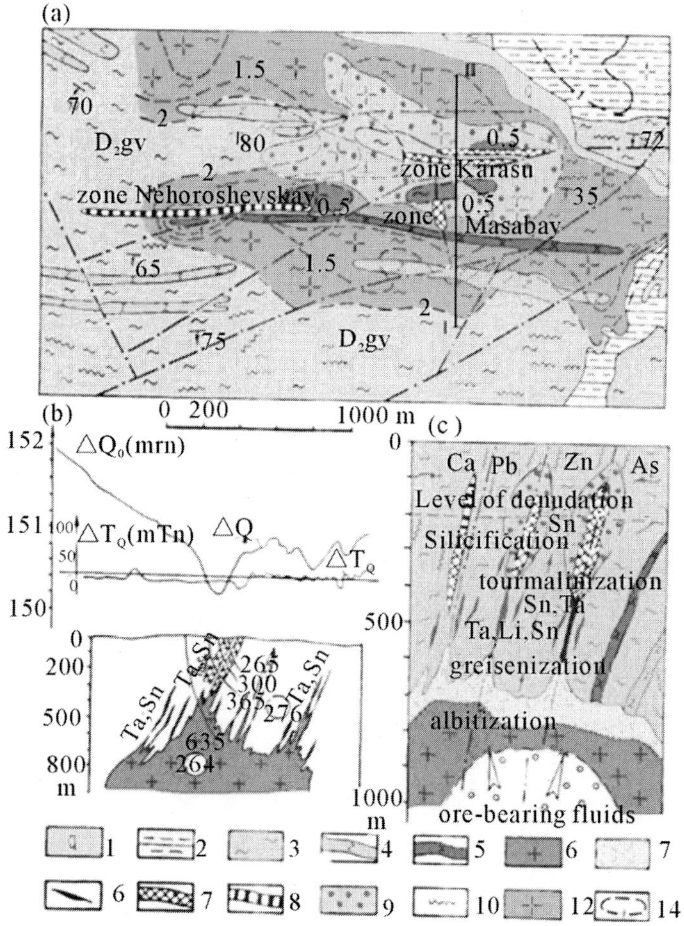

Figure 5. Geological structure of Karasu deposit. (a)—geology plan; (b)—geology-geophysical section; (c)—geologygenetic model [2]. 1—guaternary; 2—gravelly-pebbly sediments of turangy series; 3—calcic sandstones, aleurolites of kystav-kurchumskoy series (D_2gv); 4—diabase-porphyrites; 5—plagiogranite-porphyry of the kunushsky complex; 6— II phase kalba complex granites; 7—albitization and greisening zones with Sn-Ta miniralization; 8—mineralized dikes and apophysis of granites; 9—linear-stockwork zones with cassiterite and 10—sulfite mineralization; 11—development areal of stanniferous muscovite-gilbertite-guartz veins; 12—guartzledges barren; 13—concealed granites and 14—depth of their occurrence (km) according to gravity exporation.

In Kazakhstan continental edges in zones of insideplithic activization (on caledonids border of ChingizTarbagatay and hercineds of Zharma-Saur) AkbiikAkzhailay belt of granit-leukogranite and alkaline granite was located, specialized on rare-metal (W, Mo) and rareearths (Zr, Nb, TR, Mo) mineralization (deposits Shar, Zhaman-Koitas etc.). With alkaline granites (P_2-T_1) the formation of rare-earthses pegmatites and riebechitealbite metasomatite with pyrochlor-zircon mineralization are connected (deposit Upper Espe).

In an axial part of Great Altai (on the Kazakhstani border and Gorny-Altaisk continental edges) in Zaysan suturn zone with suboceanic type of earth crust Semipalatinsk-Buran-Byrgyn belt of monzonite sienite and granosienite with Zr-Ti geochemical specialization was generated. Granitoid massives of this belt (Preobrazhenska, Buran etc.) are perspective for search of zircon-ilmenite deposits in mesozoic residual soils (Karaotkel deposit), on the account of washupagain which the ilmenite looses (Satpaev placer) are formed.

The revealed laws of formation and spatial accommodation of deposits of rare metals in Kalba-Narym zone and other geological structures are the important criterion for forecasting and searching the new ores objects on East Kazakhstan territory.

4. CONCLUSIONS

Kalba-Narym is considered as allogenic cluster (Terrane), attached to the structure of the Great Altai in Hercynian collision period. The main ore-bearing structures are confined to clusters of sialitic profile with increased capacity of Meta granite layer and the earth's crust in whole. Granite Belt has a regional growth (the length of more than 500 km). Large scale of granitoids expansion emphasizes high energy and material resources for the processes of ore formation. Structural and metallogenic model of Kalba-Narym belt reflects the connection between the ore-magmatic systems with depth zones of the Earth crust and upper mantle, and therefore granitoid belt was formed as a result of long-term abyssal evolution of earth's sial zone matter.

Granite intrusions are presented to be the most orebearing, which are formed in the mobile geodynamic environment that promotes more intensive processes of ore formation in nonequilibrium PT conditions and, ultimately, formation of commercial deposits (Priirtyshsky, Belogorsky solids). On this basis, ore and magmatic systems of varied productivity degrees have been identified and described. Developed geological and genetic models and ore petrochemical typification of granitoids seem to be major factors for rare metal deposits prediction.

Recommendations to geological explorations: Involvement in the assessment reserve perspective areas, and semi-closed and covered unconsolidated sediments (Shulbinsky area, etc.). Additional appraisal of flanks and deep levels of the known ore fields and deposits (Belaya Gora, Bakenny, Upper Baymurza, Yubileiny, etc.). The search for hidden tin-tantalum mineralization in the hidden granite domes of the (Karasu, Shuruk, etc.). Development by small enterprises of small pegmatite deposits with tantalum reserves (Medvedka, Ahmetkino, Tochka, Komarovsky, etc.). Development of new prognostic and search technologies based on modern geological and geodynamic concepts of ore formation, leading ore petrological, geophysical, and mineralogical and geochemical estimation criteria, using high-precise analytical base.

Conducted metallogenic studies have shown that there are perspectives in the region to strengthen the mineral resource base for rare metal production.

5. ACKNOWLEDGEMENTS

We think the administration of the East Kazakhstan Territorial Geological Survey (Vostkaznedra) for permission to become acquainted with new results of geological exploration in the territory of the Greater Altai. This study was supported by the Science Foundation of the Ministry of Education and Science of the republic of Kazakhstan.

REFERENCES

1. G. N. Shcherba, B. A. Dyachkov, et al., "Great Altai: (Geology and Metallogeny), Book 1: Geology," Gylym, Almaty, 1998, p. 304.

2. G. N. Shcherba, Kh. A. Bespaev, B. A. Dyachkov, et al., "Great Altai: (Geology and Metallogeny). Book 2: Metallogeny," Gylym, Almaty, 1998, p. 400.

3. Kh. Bespaev, N. V. Polynsky, G. D. Granzhenko, et al., "Geology and Metallogeny of Southwest Altai: (Within the Limits of Territory of Kazakhstan and China)," Gylym, Almaty, 1997, p. 288.

4. B. A. Dyachkov, M. A. Mizernaya, N. P. Maiorova and N. A. Zimanovskaya, "Geology and Metallogeny of the East Kazakhstan Region in the Central Asian Mobil Belt," KazGeo, Existing Resource, New Horizons, Almaty, 2012, pp. 40-45.

5. B. A. Dyachkov, E. M. Sapargaliev, N. V. Polynsky, et al., "Metallogeny of the Greater Altai—The Law of Formation and the Condition of Localization of Ore Deposits," 34th International Geological Congress, Australia, 2012.

6. B. A. Dyachkov, D. V. Titov and E. M. Sapargaliev, "Ore Belts of the Greater Altai and their ore Resource Potential," Geology of Ore Deposits, Vol. 51, No. 3, 2009, pp. 197-211.

7. "Big Altai—The Unique Rare Metal-Polimetallic Province of Central Asia," Almaty, 2012, p. 407.

8. G. N. Shcherba, B. A. Dyachkov and G. P. Nakhtigal, "Metallogeni of Rudny Altai and Kalba," Science, Alma-Ata, 1984, p. 240.

9. K. R. Kovalev, Y. A. Kalinin, E. A. Naumov, F. Pirajno and A. S. Borisenko, "A Mineralogical Study of the Suzdal Sediment-Hosted Gold Deposit, Eastern Kazakhstan: Implications for Ore Genesis," Ore Geology Reviews, No. 35, 2009, pp. 186-205.

10. M. S. Rafailovich, M. A. Mizernaya and B. A. Dyachkov, "Large Gold Deposits Hosted in Black Shales: Formation Conditions and Features of Similarity," Almaty, 2011, p. 272.

11. F. A. Letnikov, "The Fludic Mode of Endogenic Processes in Continental Lithosphere and Metallogeny Problem," GEOS, 2000, pp. 204-224.

12. B. A. Dyachkov, N. P. Maiorova, G. N. Shcherba and K. A. Abdrakhmanov, "Granitoids and Ores Formation of Kalba-Narym Belt: Rudny Altai," Gylym, Almaty, 1994, p. 208.

13. "Mineral Resources of Tantalum, Niobium, Beryllium, Zirconium, and Fluorine: Geology, Economics and Technology," Ust-Kamenogorsk, 2003, p. 416.

14. B. S. Uzhkenov, G. R. Bekzhanov and L. B. Ivanov, "Rare Metals and Rar Earths of Kazakhstan," Almaty, 2011, p. 277.

15. V. Ye. Zagorsky, V. M. Makagon, B. M. Shmakin, et al., "Rare-Metal Pegmatites," Novosibirsk: Since, T.2, 1997, p. 285.

16. R. C. Wang, "Hafnon from the Koktokay # 1 Pegmatite Dyke, Altai, Xinjiang, China," 34th International Geological Congress, Australia, 2012.

17. T. Zou, H. Cao and B. Wu, "Pegmatites of Altai Province, Xinjiang," Acta Geologica Sinica, Vol. 2, No. 1, 1989, p. 45.

11

CHAPTER

Optimization of Acid Leaching of Rare-Earth Elements from Mongolian Apatite-Based Ore

Rina Kim [1], Heechan Cho [1,*], Kenneth N. Han [2], Kihong Kim [1] and Myoungwook Mun [1]

[1]Department of Energy Resources Engineering, Seoul National University, Seoul 08826, Korea

[2]Department of Materials and Metallurgical Engineering, South Dakota School of Mines and Technology, Rapid City, SD 57701-3995, USA

ABSTRACT

Optimization of the acid leaching process for Mongolian apatite-based ore containing rare-earth elements (REEs) was studied. The ore contained approximately 10% of REEs as total rare earth oxides, and the major impurities were Ca (33% as CaO) and Fe (23% as Fe_2O_3). Fe bearing minerals could be removed by passing the sample through a wet high-intensity magnetic separator before leaching. After magnetic separation, basic leaching tests were conducted to investigate the influence of the acid type and concentration, temperature, and the pH on the REE leaching level and kinetics. Hydrochloric acid was found to be the most effective leaching agent, leaching more than 90% of REEs in an hour. However, the concentrations of Ca ions in the leachate were also high, which would complicate recovery of the REEs. Therefore, to reduce the amount of Ca ions in the leachate, a two-stage leaching procedure was attempted. In stage 1, the sample was preleached using 1.0 M hydrochloric acid to dissolve Ca. In stage 2, the solid residue of stage 1 was leached using 2.0 M hydrochloric acid to dissolve REEs. Consequently, this two-stage leaching significantly reduced the Ca concentration in the final leachate without affecting the leaching levels of REEs.

Keywords: rare-earth element; acid leaching; apatite; magnetic separation; two-stage leaching; Ca removal

1. INTRODUCTION

With increasing demand for rare-earth elements (REEs) and the corresponding recent surge in their prices, many countries have begun to reevaluate old rare-earth mining prospects and explore new ones. According to a 2009 estimate by the U.S. Geological Survey, Mongolia has 31 million tonnes of rare-earth reserves (approximately 17% of the world's total), exceeded only by China [1]. In light of Chinese restrictions on REE exportation, Mongolia has become a leading contender in the current rush for REE resources [1]. At present, Mongolia plans to increase its transportation infrastructure to ship mineral resources to Russia and the Far East [1]. Therefore, Mongolian resources potentially represent a major supply of REEs for Korea.

The REE ore deposit currently under investigation is located in southern Mongolia and contains over 300,000 t of rare-earth oxides (REOs) as combined REOs according to a recent exploration campaign [2]. The average REO grade in this area is 1.36% but reaches 6.15% in a newly discovered high-grade zone. This deposit consists of sheet-like magnetite–apatite ore bodies [3].

REE ores are usually processed in two steps: physical concentration and leaching. The first step typically includes crushing the ore and physically separating the REOs from the ore. This separation process dramatically increases the percentage of REOs in the working material but is successful only if the REEs are largely concentrated in a single mineral phase such as a form of bastnäsite ($REECO_3F$) or monazite ($REEPO_4$) [4]. Unfortunately, this is not the case for the REE ore under investigation, which contains primarily apatite [$Ca_{10}(PO_4)_6(OH,F,Cl)_2$] as the major source mineral. REEs in apatite usually occur as ionic substitutions for Ca within the crystal lattice rather than in discrete mineral form [5,6,7]. Intuitively, it is possible to obtain REE concentrates by selective separation of REE-bearing apatite from the ore. Froth flotation is a common process applied to phosphate ores. Consequently, various flotation tests were conducted in an attempt to obtain REE concentrates. However, the results were not satisfactory, yielding either a low recovery and/or little REE enrichment in the concentrates. Other physical concentration methods such as gravity separation and magnetic separation were also found to be ineffective in concentrating REEs, but magnetic separation showed positive effects on impurity removal.

The extraction of REEs from ore usually involves decomposition with acidic or alkaline reagents. In the acid route, acid baking is commonly used for the major REE minerals including monazite, bastnäsite, and xenotime. The treatment involves mixing REE concentrates with sulfuric acid followed by baking at around 200 °C. The resulting cake is leached with water to dissolve REEs. Alkaline digestion is typically used for monazite. The ore is decomposed with NaOH to produce rare earth hydroxides. The resulting products are then leached using hydrochloric acid to dissolve REEs. In addition, direct decomposition by HCl is another common method used for carbonate minerals such as bastnäsite.

Several studies have examined the acid leaching of REE apatite [8,9,10,11,12]. It was reported that apatite leached well in hydrochloric or nitric acid [8,10,11], but it was very difficult to attack with sulfuric acid unless the concentration and/or temperature were very high [8,10,12]. However, owing to the complex metallurgy of REEs, there is no standard process for leaching REE-bearing minerals. The choice of decomposition method depends on the mineralogy of the REE-containing phases as well as other non-REE minerals present in the ore that often affect the effectiveness of the leaching process.

This study investigates the leaching behavior of a Mongolian REE ore in response to various acids. To reduce impurities in the leachate, pre-removal of Fe by magnetic separation and a two-stage leaching process for selective removal of Ca were studied. The specific objectives were to examine the mineralogical characteristics of the sample, determine the effects of the acid type and concentration, temperature, and the pH on the leaching efficiency, and assess the optimum conditions of the leaching process for this apatite-based ore.

2. MATERIALS AND METHODS

2.1. Mineralogical Characteristics of the Sample

The sample was collected from the southern region of Mongolia. Previous studies reported that the REE deposit consists of apatite ore bodies including apatite, barian celestite, magnetite, hematite, goethite, fluorite, gypsum, phlogopite, carbonate (mostly calcite), pyrite, and monazite-(Ce) [3,13]. The chemical composition of the sample was analyzed by X-ray fluorescence (XRF) and summarized in Table 1. The total level of REOs is approximately 10%. The most abundant REE is Ce, followed by La, Nd, Pr, and Y.

Table 1. X-ray fluorescence (XRF) analysis of the sample.

Formula	Concentration (%)	Formula	Concentration (%)
CaO	32.75	MgO	0.53
Fe_2O_3	22.91	SrO	0.47
P_2O_5	18.30	Na_2O	0.45
SiO_2	9.65	Pr_6O_{11}	0.42
CeO_2	5.80	MnO	0.29
La_2O_3	3.01	Y_2O_3	0.28
SO_3	1.80	K_2O	0.25
Nd_2O_3	1.48	ThO_2	0.06
Al_2O_3	0.87	ZnO	0.05
TiO_2	0.62	CuO	0.04

X-ray diffraction (XRD) (D/MAX-2500V/PC, Rigaku, Tokyo, Japan) was used to characterize the mineral phases in the sample. As shown in Figure 1, the main minerals were identified as apatite and silicate minerals. Minor minerals were goethite, hematite, maghemite, quartz, monazite, and so on.

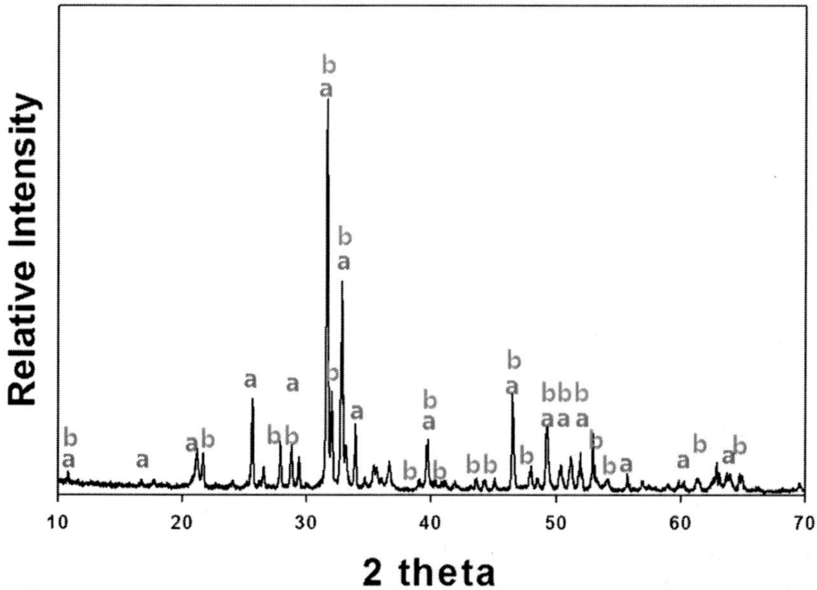

Figure 1. X-ray diffraction (XRD) analysis of the sample. (**a**) (Ce, La, Nd) hydroxyapatite, fluorapatite, chlorapatite; (**b**) Nd, Mn silicates.

Detailed mineralogical characterization of the ore was conducted by Enkhbayar et al. [14], who found that REEs were mostly associated with fluorapatite and hydroxyapatite. As shown in Figure 2, the apatite grains have hexagonal structures with a layered arrangement of outer crystals grown upon an inner layer of crystals. Electron probe microanalysis revealed that the REE (Ce_2O_3 + La_2O_3 + Nd_2O_3) content increases from the center to the edge of the specimen (9.45 wt % in mh21, 12.51 wt % in mh23, 45.57 wt % in mh24). The Si and Na contents also increased from the center to the edge of the structure. Conversely, the amounts of P and Ca decreased from the center to the edge of the sample. These trends indicate that more REE was substituted for Ca in apatite as the crystals grew. Monazite crystals were also found between the two outer layers. Therefore, it can be expected that the leaching rate would initially be high owing to the presence of a high-grade REE layer in the outer rim of the apatite particles and then decrease because monazite is less reactive than apatite [15].

Figure 2. The layered structure of apatite in the sample [14].

2.2. Experimental Procedure

Lumps of the ore sample were crushed into smaller pieces with a hammer and were crushed again to −1 mm in stages by a jaw crusher and disk mill. They were then ground to 90% of the passing size of 0.3 mm using a ball mill. Figure 3 shows the cumulative size distribution of the sample.

Because the chemical analysis indicated that the sample contained a large amount of Fe (23% Fe_2O_3), Fe-bearing minerals were removed from the sample before leaching. A mixture of water and the sample, bearing 10 wt % solid materials, was prepared and fed through a laboratory wet high-intensity magnetic separator (Eriez Series L-4 Model, Eriez, Erie, PA, USA) operating at 0.2 T. The chemical compositions of the resulting magnetic and nonmagnetic streams were analyzed using XRF.

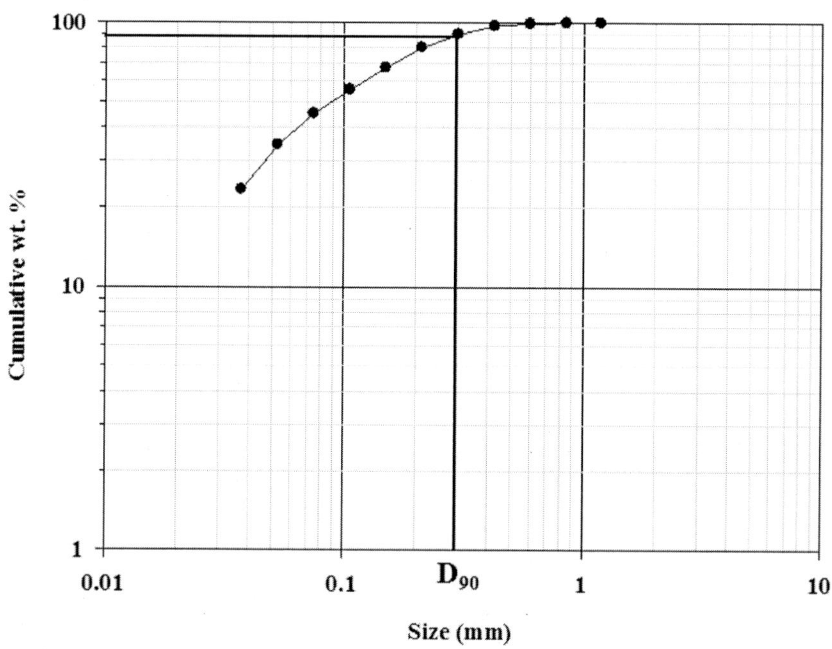

Figure 3. Cumulative size distribution of the sample after crushing and milling.

Basic leaching tests were conducted using sulfuric, hydrochloric, and nitric acids at various concentrations in the ranges of 0.1–13.0 M for sulfuric acid and 0.5–2.0 M for hydrochloric and nitric acid. All of the leaching tests were replicated 4–5 times, and average values obtained after removing significant outliers were used for analysis. Ten grams of the sample were added to a beaker containing 90 mL of the acid solution. The slurry was then agitated with an impeller-type stirrer (Daehan Scientific, Wonju, Korea). The slurry samples were collected at regular intervals and the leachate was separated from the solid particles using a filter for lower acid conditions (0.1–2.0 M acids) or a centrifuge for higher acid conditions (5.0–13.0 M acids). The separated leachate was diluted with 1.0 M hydrochloric acid to prevent precipitation of REEs and analyzed using an inductively coupled plasma optical emission spectrometer.

As the leachate contained a considerable amount of Ca, a two-stage leaching method was employed to reduce the amount of Ca in the final leachate. In stage 1, the sample was preleached using 1.0 M hydrochloric acid to dissolve Ca. In stage 2, the leachate of stage 1 was separated from the residue, and only the solid residue was leached using 2.0 M hydrochloric acid to dissolve the REEs.

3. RESULTS AND DISCUSSION

3.1. Magnetic Separation

Table 2 presents a chemical analysis of both the feed sample and the magnetic and nonmagnetic streams collected from the magnetic separator. The Fe_2O_3 content in the nonmagnetic portion decreased significantly from the feed value of 22.9% to 10.6%. In addition, the REE contents increased correspondingly, as REE-containing apatite is nonmagnetic. Some REEs were lost to the magnetic stream, but the recovery of the REEs in the nonmagnetic stream was still greater than 90%. The Fe_2O_3 content of the magnetic stream was very high, and the Fe grade was compatible with typical iron-ore concentrates. Therefore, magnetic separation before leaching offers a potential opportunity not only to reduce Fe impurities in the leachate but also to generate revenue by selling the Fe-rich byproduct to smelters.

Table 2. Result of XRF analysis of raw sample and magnetically separated samples.

Magnetic Intensity	Sample	Yield (%)	Content (wt %)					
			Fe_2O_3	CeO_2	La_2O_3	Nd_2O_3	Pr_6O_{11}	Y_2O_3
-	Raw	100	22.9	5.8	3.0	1.5	0.4	0.3
0.2 T	Magnetic	19.5	72.5	1.5	0.9	0.4	0.1	0.1
	Nonmagnetic	80.5	10.6	6.8	3.6	1.6	0.5	0.3
Recovery to Nonmagnetic (%)			37.6	95.0	94.5	94.9	96.7	96.1

3.2. Effect of Acid Concentration and Type

Figure 4 shows the change in the leaching efficiency with the sulfuric acid concentration. The experiments were run for 5 h at

20 °C. At low acid concentrations, the leaching level increased with increasing acid concentration. However, the leaching level decreased at 5.0 M. This may result from reprecipitation of REEs via calcium sulfate formation. It is known that calcium sulfate hydrates (gypsum, hemihydrate, and anhydrite) are readily formed wherever calcium and sulfate are present together in aqueous solutions [16].

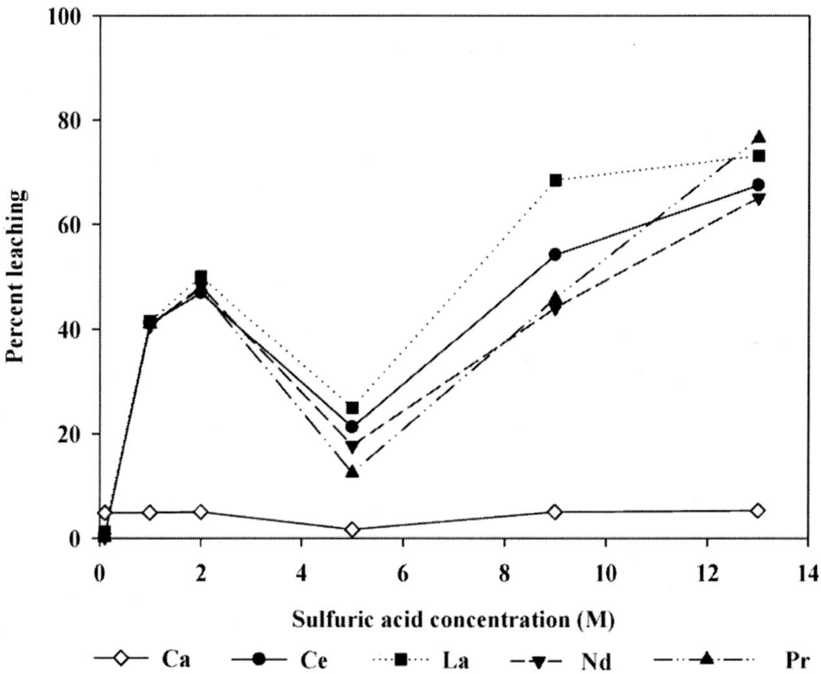

Figure 4. Leaching levels at various initial sulfuric acid concentrations at 20 °C.

Indeed, the XRD analysis of the residue after leaching shows the presence of gypsum ($CaSO_4 \cdot 2H_2O$) and anhydrite ($CaSO_4$) (Figure 5). This residue may contain REEs by isomorphous substitutions for Ca^{2+} [8,12,17,18], because Ca and REEs have similar ionic radii. To confirm the existence of REEs in the precipitated gypsum and $CaSO_4$, the leaching residue was dried in atmosphere and was washed with water. The solution phase from this washing was found to contain a significant amount of Ca and REEs (1.03 g/L Ca, 0.08 g/L Ce, 0.03 g/L La, 0.03 g/L Nd, 0.02 g/L Pr, and 0.009 g/L

Y), which indicates that the precipitated calcium sulfate hydrates contain REEs by ion substitution. The Ca concentration seems to be higher than the solubility of gypsum (2.4 g/L), but the solution may contain free calcium ions and associated calcium sulfate neutral species [$CaSO_{4(aq)}$] dissolved from anhydrite [16,19].

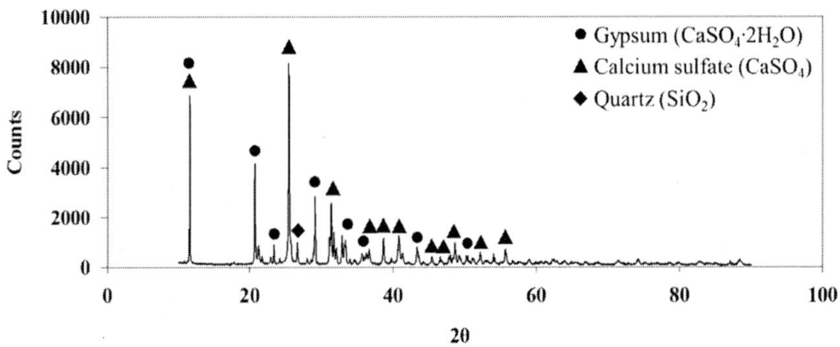

Figure 5. XRD analysis of residue from leaching with 13.0 M sulfuric acid.

On the other hand, when the concentration of sulfuric acid further increased above 5.0 M, the leaching level increased again. It can be postulated that the increase in the acidity may change the solution chemistry to limit the precipitation of calcium sulfate hydrates. To examine more closely the relationship between the leaching patterns and H_2SO_4 concentration, the Ca speciation was calculated for the Ca-SO_4-H_2O system using thermodynamic data available in the literature [20,21,22]:

$$HSO_4^- + H_2O = H_2SO_4 + OH^- \quad K_1 = \frac{[H_2SO_4][OH^-]}{[HSO_4^-]} = 3.002 \times 10^{-26}, \tag{1}$$

$$HSO_4^- + OH^- = SO_4^{2-} + H_2O \quad K_2 = \frac{[SO_4^{2-}]}{[HSO_4^-][OH^-]} = 1.034 \times 10^{12}, \tag{2}$$

$$Ca^{2+} + OH^- = Ca(OH)^+ \quad K_3 = \frac{[Ca\,(OH)^+]}{\left[Ca^{2+}\right][OH^-]} = 25.119, \tag{3}$$

$$Ca^{2+} + SO_4^{2-} = CaSO_{4\,(aq)} \quad K_4 = \frac{\left[CaSO_{4\,(aq)}\right]}{\left[Ca^{2+}\right]\left[SO_4^{2-}\right]} = 1.157 \times 10^2, \tag{4}$$

$$Ca^{2+} + 2OH^- = Ca(OH)_2 \quad K_{s1} = \left[Ca^{2+}\right][OH^-]^2 = 3.890 \times 10^{-6}, \tag{5}$$

$$Ca^{2+} + SO_4^{2-} = CaSO_{4\,(s)} \quad K_{s2} = \left[Ca^{2+}\right]\left[SO_4^{2-}\right] = 4.930 \times 10^{-5}, \tag{6}$$

$$Ca^{2+} + SO_4^{2-} + 2H_2O = CaSO_4 \cdot 2H_2O \; K_{s3} = \left[Ca^{2+}\right]\left[SO_4^{2-}\right] = 3.140 \times 10^{-5}. \tag{7}$$

There are seven species in the solution: OH^-, H_2SO_4, HSO_4^-, SO_4^{2-}, Ca^{2+}, $Ca(OH)^+$, $CaSO_{4(aq)}$. Therefore, three more equations in addition to Equations (1)–(4) are required to calculate the equilibrium composition. Two additional equations used were mass balance equations for Ca species and sulfate species. One additional equation could be the charge balance equation, but in our calculations, the OH^- concentrations were assumed to be known and were calculated from the final pH of the solutions in the actual experiment. The total Ca concentration was assumed to be 0.5 M considering the number of moles of Ca in the ore. The equilibrium composition was then calculated under different levels of total sulfate concentrations (1.0, 2.0, 5.0, 9.8, and 13.0 M). If the calculated values were greater than any of the K_{sp} values [Equations (5)–(7)], the Ca^{2+}concentration was recalculated using the Equations (5)–(7) and the calculations were repeated until the calculated values satisfied all conditions. The exact solution was obtained by an iterative method using "Solver" function in MS Excel (Microsoft, Redmond, WA, USA).

Figure 6 shows the calculation results, showing the distribution of Ca species as a function of the initial sulfuric acid concentration. In this highly acidic condition, Ca was present mainly in the form of Ca^{2+} and $CaSO_{4(aq)}$. The concentration of $CaSO_{4(aq)}$ does not change with the sulfuric acid concentration, as the solution become saturated owing to the high concentrations of Ca^{2+} and SO_4^{2-}. On the other hand, the concentration of Ca^{2+} changes with the sulfuric acid concentration, showing a minimum at 5.0 M. The final pH of the solution decreased initially with increasing the sulfuric acid concentration, but remained the same from 2.0 to 5.0 M of sulfuric acid. However, with more addition of sulfuric acid, the final pH decreased again, which accompanies the increases in the concentration of Ca^{2+}. It indicates that the concentration of Ca^{2+} species changed in a complicated manner due to the competing balance between the precipitation and dissolution of Ca, which is affected by H^+, SO_4^{2-} and Ca^{2+} in the solution. Accordingly, the total soluble Ca species varies with the sulfuric acid concentration in a pattern very similar to the leaching pattern. This result suggests that sulfuric acid may not be a good leaching agent for this ore, as the side reaction, Ca(REE) sulfate precipitation, limits the REE leaching level, which did not exceed 80%, even when a large excess of acid (13.0 M) was used.

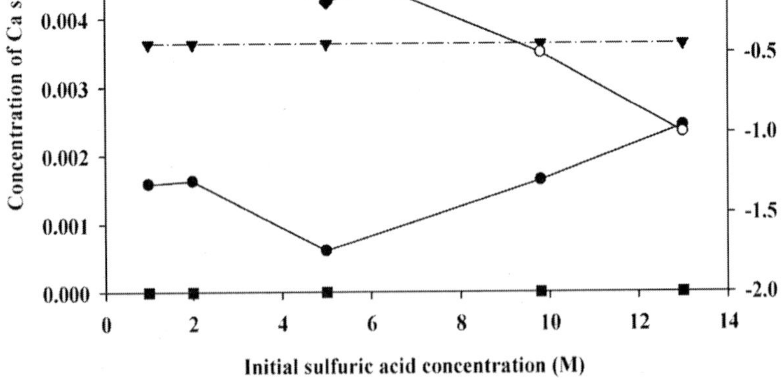

Figure 6. Ca speciation under different initial sulfuric acid concentrations at 25 °C, 1 atm. Equilibrium composition calculated by solving Equations (1)–(7).

Figure 7 shows the variation in the leaching levels with the initial concentration of hydrochloric acid. Almost no REEs were leached when the concentration was less than 1.0 M. At higher concentrations of hydrochloric acid, the leaching levels increased sharply. At an acid concentration of 2.0 M, nearly 100% of the REEs were leached. However, the amount of Ca in the leachate was also very high (27,480 mg/L), which would undoubtedly cause complications during REE recovery by either precipitation or solvent extraction.

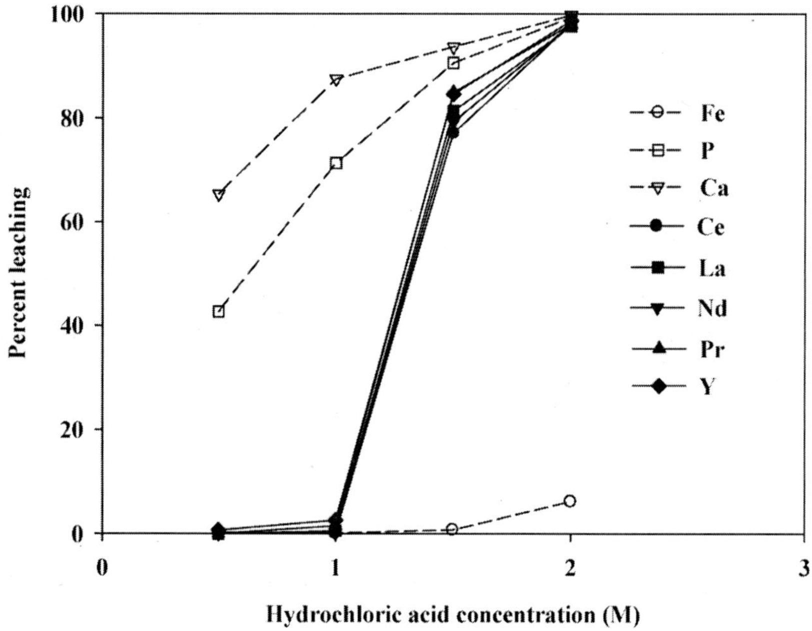

Figure 7. Leaching levels at various initial hydrochloric acid concentrations at 20 °C. Leaching time: 2 h; solid/liquid (S/L) ratio (*w*/*v*): 1:9.

Figure 8 shows the change in the leaching levels with time when 2.0 M hydrochloric acid was used as the leaching agent. More than 90% of the REEs were leached in 10 min. At 1 h, almost 100% of the REEs were leached, and the reaction reached equilibrium in 2 h. It could be noted that the rate of the leaching reaction was quite fast. Thus, it was sufficient to leach the ore at room temperature for a relatively short time.

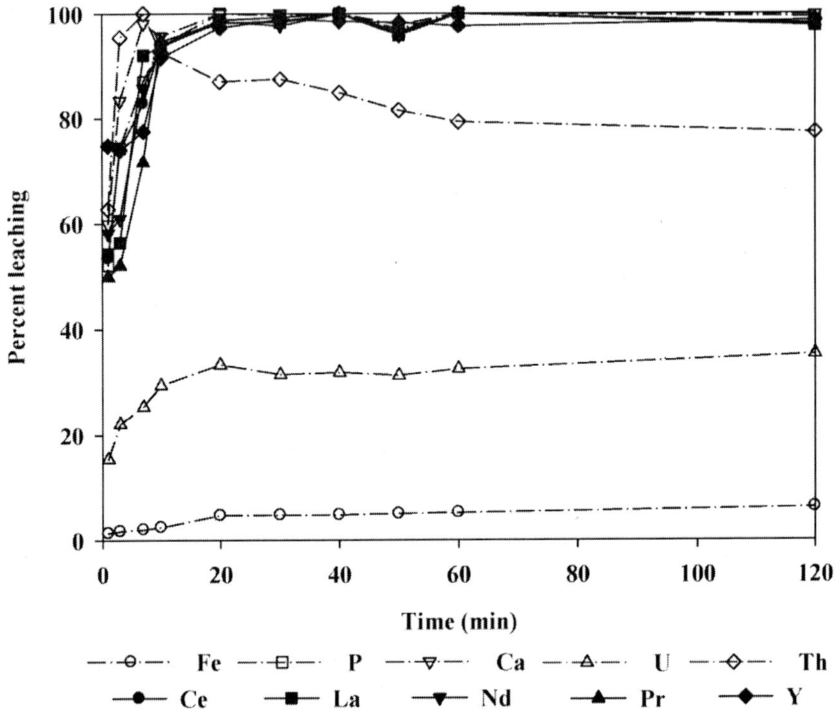

Figure 8. Change in leaching levels with time using hydrochloric acid at 20 °C. Initial acid concentration: 2.0 M; leaching time: 2 h; S/L ratio: 1:9.

Unfortunately, the radioactive elements U and Th were also leached with the REEs, although the level of radioactivity was statistically insignificant for the ore treated in this study (U: 45 mg/L, Th: 24 mg/L). Additionally, the Th leaching level decreased with time because Th precipitated as phosphate in the relatively low pH region. Thus, the radioactivity of the leaching solution could be further decreased. Note, however, that if the radioactivity of the ore was significant, the removal of radioactive elements from the leachate would require the use of separation techniques such as selective precipitation or the solvent extraction method.

Leaching tests were conducted at elevated temperatures to determine whether the leaching levels and reaction rate could be improved. Because almost 100% of the REEs were leached using

2.0 M hydrochloric acid in an hour, tests of the temperature dependence were conducted using 1.0 M hydrochloric acid at 20, 50, and 80 °C. As shown in Figure 9, the overall leaching levels for all REEs did not increase significantly; however, other impurity ions (Fe, P, Ca, and U) were leached more as the temperature increased. Therefore, it is preferable to operate with a higher concentration of hydrochloric acid at room temperature than to operate at elevated temperatures.

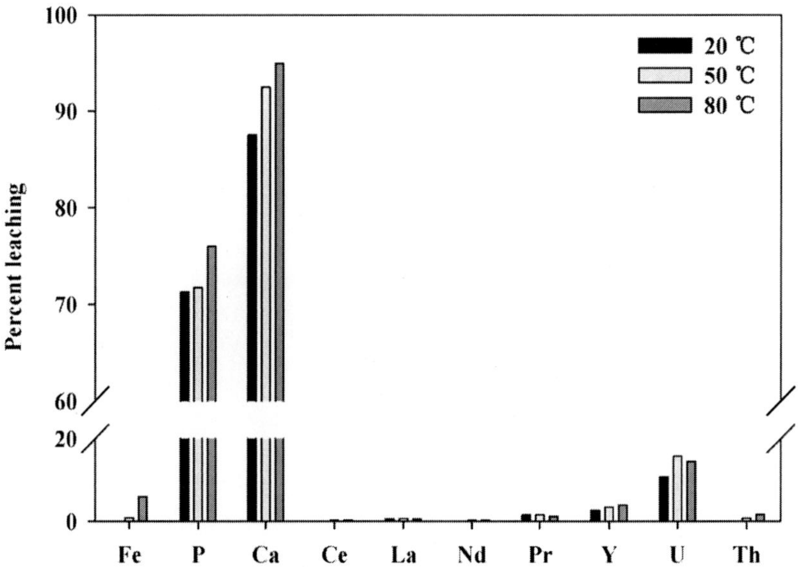

Figure 9. Comparison of final leaching levels using hydrochloric acid at 20, 50 and 80 °C. Initial acid concentration: 1.0 M; leaching time: 3 h; S/L ratio: 1:9.

Figure 10 shows the leaching levels at varying concentrations of nitric acid. No REEs, aside from Y, were leached at 1.0 M, whereas impurities such as P and Ca were leached quite well. However, as the acid concentration was increased to 2.0 M, almost 100% of the REEs were leached. These results were similar to the hydrochloric acid case. Because nitric acid is generally more expensive than hydrochloric acid, it was not considered in later sections. Consequently, hydrochloric acid was chosen as the optimum leaching agent for this apatite REE ore, and the effects of other variables on the leaching efficiency were investigated using hydrochloric acid in later experiments.

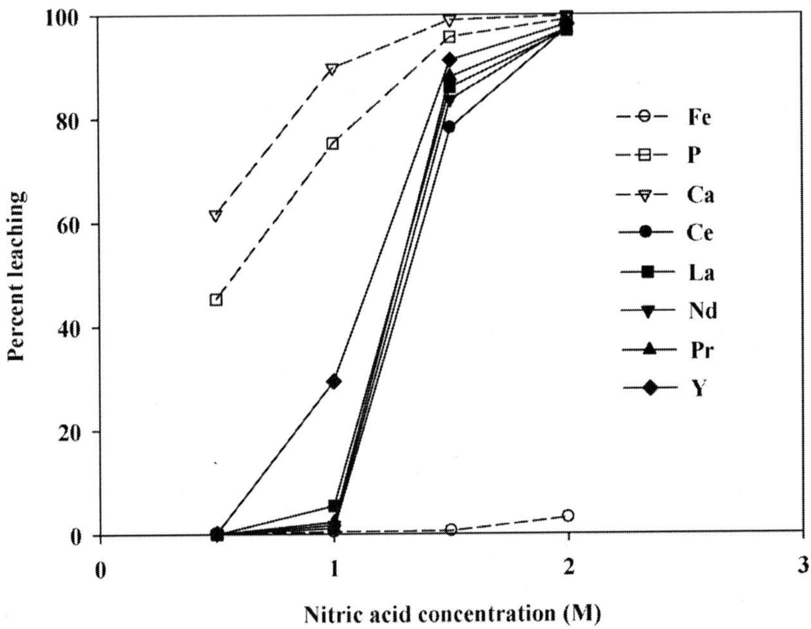

Figure 10. Leaching levels at various initial nitric acid concentrations at 20 °C. Leaching time: 2 h; S/L ratio: 1:9.

3.3. Two-Stage Leaching

As mentioned previously, the leachate obtained using 2.0 M hydrochloric acid contained a significant amount of Ca, as the major mineral phase of the sample is apatite. Therefore, a two-stage leaching process was attempted to reduce the amount of Ca in the leachate.

3.3.1. Stage 1 Leaching

In stage 1, 1.0 M hydrochloric acid was used to dissolve the apatite. Figure 11 shows the change in the leaching levels of P and Ca ions together with the REEs over time. Within the first 10 min of leaching, the REE leaching levels were higher than those of P and Ca. This may result from the mineralogical structure of the ore. As mentioned earlier, the REE concentration was higher at the outer edge of the apatite grains than at the center. The monazite grains, which bore a much higher percentage of REEs, were also present at the surface of the apatite. Conversely, the amounts of P and Ca increase gradually from the edge to the center of the apatite grains. Thus, a high initial leaching level of REEs is expected, as the leaching solution reacts first with the outer layer of the solid particles. After 10 min, the REE levels in the leachate decreased. It seemed that this could be attributed to precipitation of REEs as phosphates.

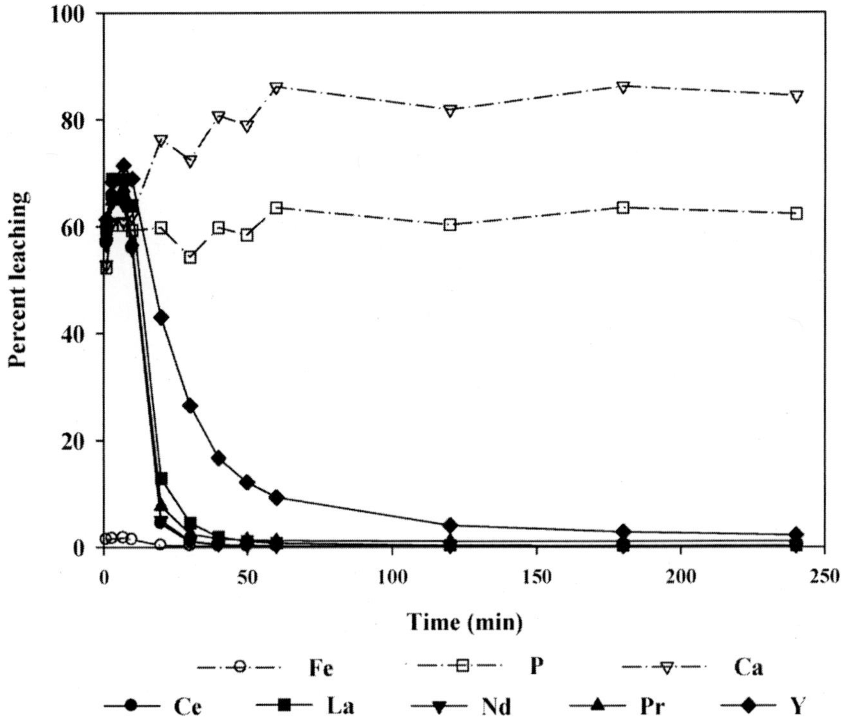

Figure 11. Change in leaching levels over the course of stage 1 using hydrochloric acid at 20 °C. Initial acid concentration: 1.0 M; leaching time: 4 h; S/L ratio: 1:9.

To analyze the leaching behavior more specifically during stage 1, a thermodynamic analysis was performed to calculate the leaching pH values for apatite, REE phosphates, and Ca phosphate using thermodynamic data available in the literature

[22,23,24,25,26,27]. It was assumed that the concentration of the substance reaches 1.0 mol/L for apatite and Ca phosphate, and 0.1 mol/L for REE phosphates. The leaching pH values were calculated on the basis of the chemical reactions given in Equations (8)–(10). The results are summarized in Table 3 and Table 4. Fluorapatite, hydroxyapatite, and chlorapatite leach at pH values of 1.0, 3.2, and 2.3, respectively. Therefore, in the early stages of leaching with 1.0 M hydrochloric acid, all three forms of apatite could be leached from the ore because the pH of the acid solution was 0:

$$Ca_{10}(PO_4)_6X_2 + 20H^+ = 10Ca^{2+} + 6H_3PO_4 + 2HX \ (X = F, OH, Cl), \quad (8)$$

$$Ca_3(PO_4)_2 + 6H^+ = 3Ca^{2+} + 2H_3PO_4, \quad (9)$$

$$REEPO_4 + 3H^+ = REE^{3+} + H_3PO_4. \quad (10)$$

During leaching, it was found that the pH increased from 0 to approximately 1.5. In this range, the dissolved REEs and phosphate ions can react to form REE phosphates, as the leaching pH for REE phosphates is less than or around 0 (Table 4). In contrast, Ca does not precipitate because both apatite and Ca phosphate are still leachable in this pH region, as shown inTable 3. Therefore, Ca dissolution continues during this stage, whereas the amount of REEs in the liquid phase decreases as the REE phosphates precipitate.

Table 3. Thermodynamic data and leaching pH of apatite and calcium phosphate according to Equations (8) and (9), respectively ($\Delta G°_f$ of H^+: 0 kJ/mol; Ca^{2+}: −553.58 kJ/mol; H_3PO_4: −1142.54 kJ/mol).

Apatite	$\Delta G°_f$(kJ/mol)			Leaching pH
	Apatite	HX[1]	$\Delta G°_{rxn}$	
$Ca_{10}(PO_4)_6F_2$ Fluorapatite	−12,885.9	−296.82	−98.76	1.00
$Ca_{10}(PO_4)_6(OH)_2$ Hydroxyapatite	−12,503.5	−237.18	−361.93	3.24
$Ca_{10}(PO_4)_6Cl_2$ Chlorapatite	−12,403.0	−131.23	−250.50	2.33
$Ca_3(PO_4)_2$	−3884.82	-	−61.00	1.84

[1] X = F, OH, Cl for fluor, hydroxy and chlorapatite, respectively.

Table 4. Thermodynamic data and leaching pH of REE phosphates according to Equation (10) ($\Delta G°_f$ of H^+: 0 kJ/mol; H_3PO_4: −1142.54 kJ/mol).

REE Phosphate	$\Delta G_f°$ (kJ/mol)			Leaching pH
	$REEPO_4$	REE^{3+}	$\Delta G°_{rxn}$	
YPO_4	−1846.96	−686.92	17.50	−0.36
$LaPO_4$	−1898.43	−730.46	25.43	−0.82
$CePO_4$	−1827.32	−677.33	7.45	0.23
$PrPO_4$	−1880.33	−712.88	24.92	−0.79
$NdPO_4$	−1870.92	−704.09	24.29	−0.75

In Figure 7 and Figure 11, it can be noted that the percentage of dissolved P is always less than that of Ca for 1.0 M hydrochloric acid. This also implies that the dissolved P is consumed as the REE phosphates form. Otherwise, the leaching levels of P, REEs, and Ca would be similar over time. Table 5 presents the hypothetical P leaching levels if no precipitation occurred. In these calculations, it was assumed that the REEs disappeared from the solution according to the reaction REE3++PO3−4=REEPO4; i.e., moles of phosphate ions and REEs were removed from the solution in a 1:1 ratio as precipitation progressed. When the number of moles of phosphate ions consumed by the precipitation reaction is calculated and added to the experimentally measured phosphate level, the calculated leaching levels of P are very close to those of Ca throughout the process. This confirms that the REE phosphates precipitate out of the solution during this stage.

Indeed, some REE phosphates were identified by XRD analysis of the residue from stage 1, as shown in Figure 12. This indicated that it would be possible to dissolve Ca and P selectively from the ore while keeping the REEs in the solid residue by using 1.0 M hydrochloric acid for a reasonable leaching time, approximately 3–4 h.

Table 5. Comparison of P and Ca leaching level during stage 1 leaching.

Leach Time (min)	P Leaching Level		Ca Leaching Level
	Actual	Re-Calculated	
1	52.25	52.26	52.93
3	60.26	60.28	60.86
7	60.29	60.31	61.01
10	59.29	60.23	62.07
20	59.83	74.53	76.31
30	54.33	69.30	72.41
40	59.83	76.87	80.74
50	58.44	75.20	79.01
60	63.50	81.83	86.16
120	60.24	77.70	81.80
180	63.42	81.83	86.16
240	62.21	80.25	84.37

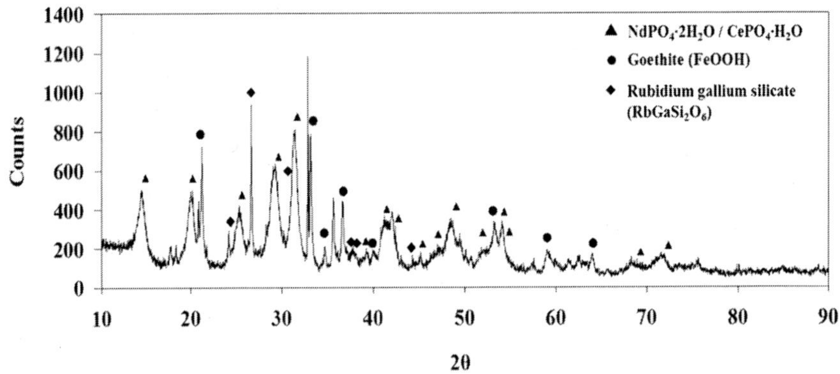

Figure 12. XRD analysis of the residue from stage 1, showing the presence of REE phosphates.

3.3.2. Stage 2 Leaching

The solid residue from stage 1 was subjected to stage 2 leaching using 2.0 M hydrochloric acid as a leaching agent. Figure 13 shows that the leaching of REEs reaches nearly 100% in a short time. Table 6 compares the leaching levels of REEs, calculated as the ratio of the amount of REEs in the leachate after 120 min of leaching to that in the raw sample, with those obtained by one-stage leaching under the same conditions. The amounts of the major impurities (Ca, P, and Fe) in the final leachate are also shown. The leaching levels of the REEs were not affected by the use of two-stage leaching, indicating that only the impurities were selectively leached during the first stage of leaching. As a consequence, the amounts of Ca and P in the final leachate were considerably reduced from 27,480 and 11,260 mg/L to 5030 and 4400 mg/L, respectively.

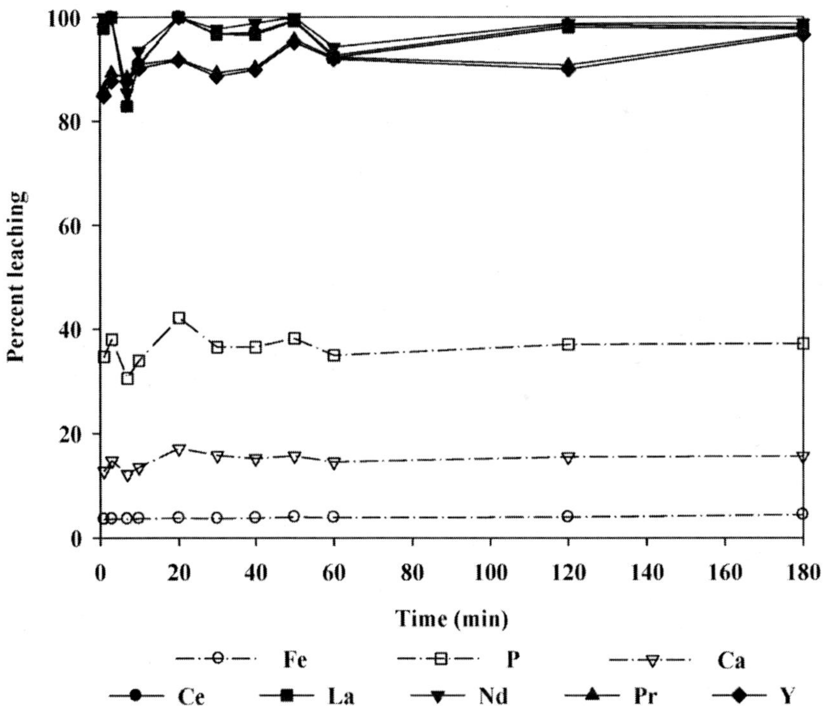

Figure 13. Leaching levels for various elements over the course of stage 2 using hydrochloric acid. Initial acid concentration: 2.0 M; leaching time: 3 h; S/L ratio: 1:9.

Table 6. Comparison of REE leaching levels in the leachate after 120 min and the amount of some impurities in the final leachate between one and two-stage processes.

Type of Leaching	REE Leach Level (%)					Amount of Impurity (mg/L)		
	Ce	La	Nd	Pr	Y	Ca	P	Fe
One-stage	98.1	97.7	98.0	98.0	98.7	27,480	11,260	190
Two-stage	98.4	97.9	98.8	90.7	90.0	5030	4400	147

3.4. Optimal Processing Route

On the basis of the experimental results of this study, we propose a route for optimal processing and leaching of REEs from this ore, as shown in Figure 14. In this process, the ore is crushed and ground to 90% passing 0.3 mm. Then, magnetic separation is applied to separate Fe bearing minerals, which results in 60% removal of Fe with a 5%–6% loss of REEs. The Fe grade of the magnetic product is very high (over 70%), so it can be sold to the steel-making industry. The nonmagnetic stream is then subject to a two-stage leaching process. In stage 1, the ore is leached with 1.0 M HCl, where approximately 90% of the Ca is removed with no loss of REEs. The residue from stage 1 leaching is then subject to stage 2 leaching with 2.0 M HCl, where most of the REEs are leached out in an hour. The final recovery of REEs after leaching is around 92%–94% for each REE—a considerably higher amount than that yield by other current processing practices. Table 7 summarizes the metallurgical balance of the proposed processing flow.

Table 7. Metallurgical balance of the optimal process.

Process	Sample	Yield (Wt %)	Elements (%)							Distribution (%)						
			Ca	Fe	Ce	La	Nd	Pr	Y	Ca	Fe	Ce	La	Nd	Pr	Y
Magnetic separation	Magnetic	19.51	8.08	72.52	1.48	0.87	0.36	0.07	0.05	5.04	62.37	5.04	5.49	5.09	3.30	3.89
	Nonmagnetic	80.49	36.93	10.61	6.76	3.64	1.65	0.53	0.32	94.96	37.63	94.96	94.51	94.91	96.70	96.11
1st stage leaching	Leachate	46.06	54.44	0.02	0.01	0.01	0.00	0.01	0.01	80.12	0.05	0.06	0.16	0.06	1.00	2.11
	Residue	34.42	13.50	24.78	15.80	8.51	3.85	1.22	0.73	14.84	37.59	94.90	94.35	94.85	95.70	94.00
2nd stage leaching	Leachate	17.14	27.01	2.20	31.11	16.70	7.64	2.40	1.45	14.79	1.66	93.05	92.24	93.70	93.71	92.76
	Residue	17.28	0.09	47.18	0.62	0.38	0.09	0.05	0.02	0.05	35.92	1.85	2.10	1.15	1.99	1.24
Calculated feed		100.0	31.30	22.69	5.73	3.10	1.40	0.44	0.27	100.00	100.00	100.00	100.00	100.00	100.00	100.00
Measured feed		100.0	32.75	22.91	5.80	3.01	1.48	0.42	0.28	-	-	-	-	-	-	-

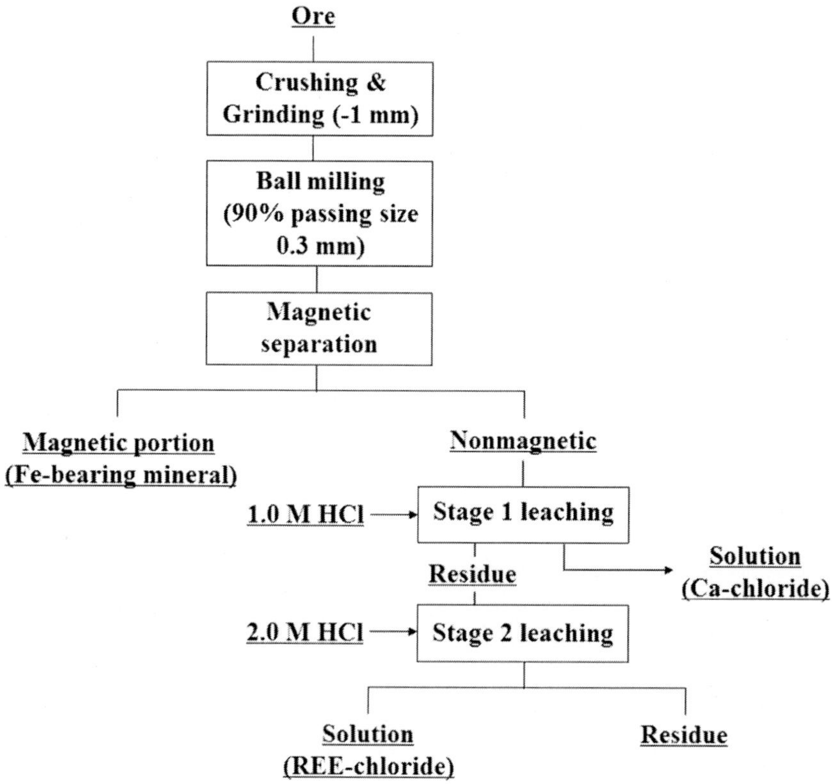

Figure 14. Optimal processing of the ore.

4. CONCLUSIONS

The REE ore deposit currently under investigation can become a new major source of REEs judging from its magnitude and grade. Furthermore, test results show that this ore can be processed by a rather simple route with a recovery of all REEs of over 90% REEs. It seems that physical separation of REE-containing minerals may not be easy. However, removal of Fe-bearing minerals before leaching is possible using magnetic separation with minimal loss of REEs. This would reduce the operation cost of the leaching process because much of the acid could be consumed by Fe minerals in the ore. At the same time, additional profit can be realized by the sale of Fe concentrate as the magnetic product has a smelter grade.

Leaching tests reveal that REEs can be effectively extracted using HCl at ambient temperature and pressure. The U and Th content of the leachate was not at a significant level that demands additional treatment. However, the Ca level was very high when direct leaching with HCl was performed. Consequently, a two-stage leaching process was developed. In stage 1, leaching is conducted under a mild condition (1.0 M HCl) to elute Ca from the ore. In stage 2 leaching, 90%–100% of the REEs were recovered using 2.0 M HCl. The final recovery of REEs after leaching is very high around 92%–94% for each REE—a considerably higher amount than that yield by other current processing practices.

ACKNOWLEDGMENTS

This work was supported by a grant from the Korea Institute of Energy Technology Evaluation and Planning (KETEP) funded by the Ministry of Trade, Industry, and Energy (MOTIE) of the Korean government (No. 20122010300041).

AUTHOR CONTRIBUTIONS

Rina Kim performed the experiments, and analyzed and interpreted the data. Heechan Cho was the Principal Investigator of this work and edited the paper. Kenneth N. Han contributed to interpretation of the data. Kihong Kim and Myoungwook Mun performed magnetic separation tests and helped to analyze the data.

CONFLICTS OF INTEREST

The authors declare no conflict of interest.

REFERENCES

1. Daly, J.C.K. Mongolia's rare earth reserves draw foreign investor interest. *Centr. Asia-Caucasus Anal.* 2011, *13*, 6–8.

2. Altangerel, T. Rare Earth Deposits in Mongolia. Available online: http://criticalmetalsmeeting.com/download/CM-2013_Finalized_program.pdf (accessed on 30 June 2016).

3. Sanematsu, K.; Kon, Y.; Hirando, H.; Watanabe, Y. Mineralogy and Geochemistry of the Mushgai Kudag REE Deposit, Mongolia. In *Geological Society of America Abstracts with Programs*; The Geological Society of America: Denver, CO, USA, 2010; Volume 42, p. 335.

4. Chi, R.; Tian, J.; Zhu, G.; Wu, Y.; Li, S.; Wang, C.; Zhou, Z.A. Kinetics of rare earth leaching from a mangnese-removed weathered rare-earth mud in hydrochloric acid solutions. *Separ. Sci. Technol.* **2006**, *41*, 1099–1113.

5. Zielinski, S.; Szczepanik, A.; Buca, M. Recovery of lanthanides from Kola Apatite in phosphoric acid manufacture. *J. Chem. Technol. Biotechnol.* **1993**, *56*, 355–360.

6. Jones, A.P.; Wall, F.; Williams, C.T. *Rare Earth Minerals—Chemistry, Origin and Ore Deposits*; Chapman & Hall: London, UK, 1996.

7. Martin, P.; Carlot, G.; Chevarier, A.; Den-Auwer, C.; Panczer, G. Mechanisms involved in thermal diffusion of rare earth elements in apatite. *J. Nucl. Mater.* **1999**, *275*, 268–276.

8. Jorjani, E.; Bagherieh, A.H.; Chelgani, S.C. Rare earth elements leaching from Chadormalu apatite concentrate: Laboratory studies and regression predictions. *Korean J. Chem. Eng.* **2011**, *28*, 557–562.

9. Aly, M.M.; Mohammed, N.A. Recovery of lanthanides from Abu Tartur Phosphate Rock, Egypt. *Hydrometallurgy* **1999**,*52*, 199–206.

10. Sandström, Å.; Fredriksson, A. Apatite for extraction—Leaching of Kiirunavaara Apatite for simultaneous production of fertilizers and REE. In Proceedings of the

26th International Mineral Processing Congress, IMPC 2012: Innovative Processing for Sustainable Growth, New Delhi, India, 24–28 September 2012; pp. 4707–4714.

11. Pereira, F.; Bilal, E. Phosphoric acid extraction and rare earth recovery from apatites of the Brazilian Phosphatic Ores.*Romanian J. Mineral. Depos.* **2012**, *85*, 49–52.

12. Habashi, F. The recovery of the lanthanides from Phosphate Rock. *J. Chem. Technol. Biotechnol. Chem. Technol.* **1985**, *35*, 5–14.

13. Garamjav, D.; Jargalan, S. Introduction to REE metallogeny of the Mongolia. *Mong. Geosci.* **2009**, *35*, 73–76.

14. Enkhbayar, D.; Seo, J.; Choi, S.; Lee, Y.; Batmunkh, E. Mineral chemistry of REE-rich apatite and sulfur-rich monazite from the Mushgai Khudag, Alkaline Volcanic-Plutonic Complex, South Mongolia. *Int. J. Geosci.* **2016**, *7*, 20–31.

15. Peelman, S.; Sun, Z.H.I.; Sietsma, J.; Yang, Y. Hydrometallurgical extraction of rare earth elements from low grade mine tailings. In *Rare Metal Technology 2016*; Alan, S., Kim, H., Neelameggham, N.R., Ouchi, T., Oosterhof, H., Eds.; John Wiley & Sons, Inc.: Hoboken, NJ, USA, 2016.

16. Azimi, G.; Papangelakis, V.G. The Solubility of gypsum and anhydrite in simulated laterite pressure acid leach solutions up to 250 °C. *Hydrometallurgy* **2010**, *102*, 1–13.

17. Todorovsky, D.S.; Milanova, M.M.; Minkova, N.L.; Balarev, C. Solubility of some lanthanide sulfate in polycomponent systems containing H_2SO_4. *Monatsh. Chem.* **1993**, *124*, 673–679.

18. Wang, L.; Long, Z.; Huang, X.; Yu, Y.; Cui, D.; Zhang, G. Recovery of rare earths from wet-process phosphoric acid.*Hydrometallurgy* **2010**, *101*, 41–47.

19. Gangolli, S. *The Dictionary of Substances and Their Effects: C*; Royal Society of Chemistry: London, UK, 1999.

20. Martell, E.A.; Smith, R.M. *Critical Stability Constants Volume 4*; Plenum Press: New York, NY, USA, 1982.

21. Lide, D.R. *CRC Handbook of Chemistry and Physics*, 83rd ed.; CRC Press: Cleveland, OH, USA, 2002.

22. Kim, E.; Osseo-Asare, K. Aqueous stability of Thorium and rare earth metals in monazite hydrometallurgy: Eh-pH diagrams for the systems Th-, Ce-, La-, Nd- (PO_4)-(SO_4)-H_2O at 25 °C. *Hydrometallurgy* **2012**, *113–114*, 67–78.

23. Firsching, F.H.; Brune, S.N. Solubility products of the trivalent rare-earth phosphates. *J. Chem. Eng. Data* **1991**, *36*, 93–95.

24. Oxtoby, D.W.; Gillis, H.P.; Nachtrieb, N.H. *Principles of Modern Chemistry*, 5th ed.; Thomson: Pacific Grove, CA, USA, 2002.

25. La Iglesia, A. Estimating the thermodynamic properties of phosphate minerals at high and low temperature from the sum of constituent units. *Estud. Geol.* **2009**, *65*, 109–119.

26. Han, K.N.; Kellar, J.J.; Cross, W.M.; Safarzadeh, S. Opportunities and challenges for treating rare-earth elements.*Geosyst. Eng.* **2014**, *17*, 178–194.

27. Kim, R.; Cho, H.C.; Han, K.N. Behavior of anions in association with metal ions under hydrometallurgical environments Part I—OH^- effect on various cations. *Miner. Metall. Proc.* **2014**, *31*, 40–47.

12

CHAPTER

Separation and Recycling for Rare Earth Elements by Homogeneous Liquid-Liquid Extraction (HoLLE) Using a pH-Responsive Fluorine-Based Surfactant

Shotaro Saito [1], Osamu Ohno [1], Shukuro Igarashi [1,*], Takeshi Kato [2] and Hitoshi Yamaguchi [3]

[1]*Department of Biomolecular Functional Engineering, College of Engineering, Ibaraki University, 4-12-1 Nakanarusawa-cho, Hitachi, Ibaraki 316-8511, Japan*

[2]*Industrial Technology Institute of Ibaraki Prefecture, 3781-1 Nagaoka, Ibaraki-machi, Ibaraki-gun, Ibaraki 311-3195, Japan*

[3]*Research Center for Strategic Materials, National Institute for Material Science, 1-2-1 Sengen, Tsukuba, Ibaraki 305-0047, Japan*

ABSTRACT

A selective separation and recycling system for metal ions was developed by homogeneous liquid-liquid extraction (HoLLE) using a fluorosurfactant. Sixty-two different elemental ions (e.g., Ag, Al, As, Au, B, Ba, Be, Bi, Ca, Cd, Ce, Co, Cr, Cu, Dy, Er, Eu, Fe, Ga, Gd, Ge, Hf, Hg, Ho, In, Ir, La, Lu, Mg, Mn, Mo, Nb, Nd, Ni, Os, P, Pb, Pd, Pr, Pt, Re, Rh, Ru, Sb, Sc, Se, Si, Sm, Sn, Sr, Ta, Tb, Te, Ti, Tl, Tm, V, W, Y, Yb, Zn, and Zr) were examined. By changing pH from a neutral or alkaline solution ($pH \geq 6.5$) to that of an acidic solution ($pH < 4.0$), gallium, zirconium, palladium, silver, platinum, and rare earth elements were extracted at >90% efficiency into a sedimented

Zonyl FSA®($CF_3(CF_2)_n(CH_2)_2S(CH_2)_2COOH$, n = 6–8) liquid phase. Moreover, all rare earth elements were obtained with superior extraction and stripping percentages. In the recycling of rare earth elements, the sedimented phase was maintained using a filter along with a mixed solution of THF and 1 M sodium hydroxide aqueous solution. The Zonyl FSA® was filtrated and the rare earth elements were recovered on the filter as a hydroxide. Furthermore, the filtrated Zonyl FSA was reusable by conditioning the subject pH.

Keywords: homogeneous liquid-liquid extraction; HoLLE; phase separation phenomenon; fluorosurfactant; ion association; rare earth elements; separation and recycling; hydrometallurgy

1. INTRODUCTION

Many kinds of rare earth elements such as neodymium, europium, and gadolinium have been recently used in the development of new materials. However, the production of scandium, yttrium, and lanthanoids, collectively referred to as rare earth elements, is not yet widespread across the world. Hence, the potential production of these materials via waste recycling processes has become increasingly important; however, the costs and efficiencies of these recycling processes pose inherent problems. The mutual metal separation method is generally classified into wet and dry methods. Here, the wet method has been employed for reasons such as separation accuracy and energy savings. This typical method utilizes a solvent extraction process which uses solvents that are immiscible with each other. In addition to this process, various separating agents and methods for metal separation have also been recently developed. These include solvent extraction using ionic liquids [1,2], phase separation extraction using stimuli-responsive polymers [3,4,5,6], and solid-phase extraction using the adsorption effect of biological materials [7,8,9]. Among these, it is especially noteworthy that in 1988, Igarashi and Yotsuyanagi discovered the pH-dependent phase separation phenomenon, resulting from the addition of acid to a perfluorooctanoic acid (PFOA) aqueous solution containing trace amounts of acetone [10]. An oily, small secondary phase is resultantly produced, with the target substance ultimately being extracted from therein.

Furthermore, the formation of a stable liquid phase which consists of $PFOA^-$ and quaternary salt (Q^+ such as tetrabutylammonium ion (TBA^+)) was subsequently reported in the following year (1989) [11,12]. The produced liquid phase is stable in water and air; hence, it is known to be the same type as the early ionic liquid reported in 1992 [13] because the liquid is produced at room temperature by ionic associations of the subject cationic and anionic organic compounds (e.g., melting point; −2.3 °C in $PFOA^-/TBA^+$[14]). This formation reaction of an ionic liquid is applied at the same time as the homogeneous liquid-liquid extraction method (HoLLE, type I in Figure 1). Many research papers have been published from the viewpoint of microextraction methods for trace components [15,16,17,18,19,20]. Hence, HoLLE, a separation method that uses these characteristics effectively, was conceived. In HoLLE, because an aqueous phase and an organic phase are in a homogeneous state at the starting point of extraction, a mechanically vigorous agitation for increasing the contact interface between the aqueous and organic phases in the solvent extraction process is not necessary. This separation method is the extraction to the water-immiscible sedimented phase for the target substance by providing stimuli such as changes in pH, temperature, and light. Some separation and concentration methods for metal chelates using surfactants such as PFOA and slight amounts of water-miscible organic solvents (such as acetone, tetrahydrofuran, *etc.*) have been reported [15,16,17,19]. In addition, a concentration method for rare earth elements using phosphoric acid and di(2-ethylhexyl)ester (D2EHPA) as an extracting agent in $PFOA^-/TBA^+$ has also been reported [21].

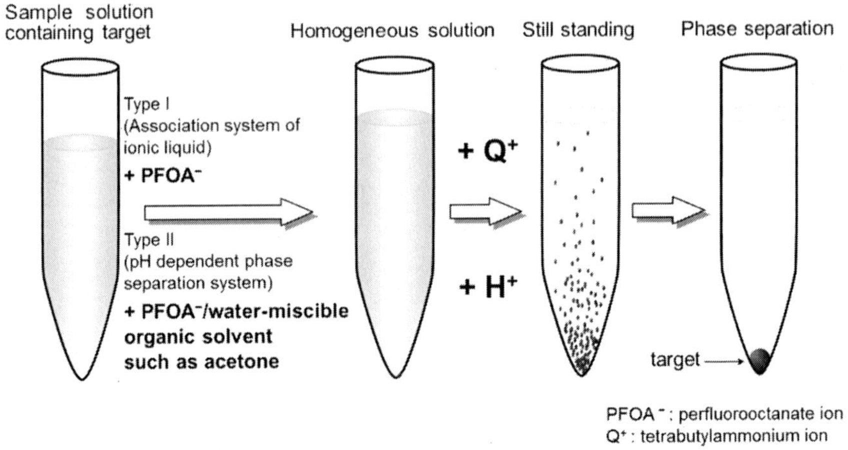

Figure 1. Scheme of homogeneous liquid-liquid extraction method (HoLLE).

In this study, Zonyl FSA®, which is an alternative substance of $PFOA^-$, was used as a phase separation agent in HoLLE. Zonyl FSA® is a 3-[2-(perfluoroalkyl)ethylthio]propanoic acid ($CF_3(CF_2)_n(CH_2)_2S(CH_2)_2COOH$, n = 6–8), with an acid dissociation constant (pK_a) of approximately 6 [22]. Therefore, this compound displays weak acidity because a spacer of some methylene groups is introduced between the fluorocarbon chain and the hydrophilic group, as compared to the strongly acidic PFOA. Applications of the separation and concentration of chlorophyll-a [22], silver complex [23], copper complex [24], indium complexes [25], and aluminum/titanium with 1,10-phenanthroline complexes [26] have been reported by HoLLE using this phase-separating agent.

In this paper, the extraction for 62 kinds of elements were examined and a high-efficiency separation and recycling system of rare earth elements using Zonyl FSA® as a phase-separating agent based on HoLLE, without the use of a chelating reagent, is described.

2. EXPERIMENTAL SECTION

2.1. Reagents

Zonyl FSA® was purchased from DuPont (Wilmington, DE, USA). Sixty-two types of elemental ions (Ag, Al, As, Au, B, Ba, Be, Bi, Ca, Cd, Ce, Co, Cr, Cu, Dy, Er, Eu, Fe, Ga, Gd, Ge, Hf, Hg, Ho, In, Ir, La, Lu, Mg, Mn, Mo, Nb, Nd, Ni, Os, P, Pb, Pd, Pr, Pt, Re, Rh, Ru, Sb, Sc, Se, Si, Sm, Sn, Sr, Ta, Tb, Te, Ti, Tl, Tm, V, W, Y, Yb, Zn, and Zr) were selected and their predetermined concentrations were prepared by diluting 1000-ppm standard solutions (received from Kanto Chemicals, Tokyo, Japan) with distilled water. Concretely, Ag(I), Al(III), Ba(II), Be(II), Bi(III), Ca(II), Cd(II), Ce(III), Co(II), Cr(VI), Cu(II), Dy(III), Er(III), Eu(III), Fe(II), Ga(III), Gd(III), Hg(II), Ho(III), In(III), La(III), Lu(III), Mg(II), Mn(II), Nd(III), Ni(II), Pb(II), Pd(II), Pr(III), Sc(III), Se(IV), Sm(III), Sr(II), Tb(III), Tl(I), Tm(III), V(V), Y(III), Yb(III), Zn(II), and Zr(IV) were 0.01–1 M of nitric acid aqueous solution. As(III), Au(III), Ir(IV), Mo(VI), Os(IV), Pt(II), Rh(III), Ru(III), Sb(III), Sn(II), and Te(IV) were 0.4–6 M of hydrochloric acid aqueous solution. Hf(IV), Nb(V), and Ta(V) were 1–3 M of hydrofluoric acid aqueous solution. Ge(IV) and Si(IV) were 0.2–0.4 M potassium hydroxide solution. Ti(IV) was 2 M sulfuric acid aqueous solution. Each of B(III), P(V), Re(VII), and W(VI) were aqueous solution of oxo acid salts containing some ammonium ion. All element solutions were diluted with distilled water to 100 fold. Tetrahydrofuran (THF), acetic acid, sodium acetate, sodium hydroxide, acetone, and all other reagents used were of commercially available analytical grade unless otherwise specified. The utilized filter (for all cases) was a Merck Millipore hydrophilic PTFE filter JAWP02500 (Omnipore, pore size: 1.0 μm, Billerica, MA, USA).

2.2. Apparatuses

The following apparatuses were used: pH meter = Model F-51, manufactured by Horiba Ltd. (Kyoto, Japan); centrifugal separator = Model LC-100, manufactured by TOMY SEIKO Co., Ltd. (Tokyo, Japan); vacuum pump = Model MDA-015, manufactured by ULVAC (Kanagawa, Japan); ICP optical emission spectrometer = Model iCAP6300, manufactured by Thermo Fisher Scientific (Waltham, MA, USA).

2.3. Experimental Procedure

2.3.1. Extraction Characteristics of Elements by HoLLE with Zonyl FSA®

The following were added into a glass vial: 0.5 mL of elemental mixed solution (each element concentration: 10 ppm), 1.0 mL of THF, and 1.0 mL of Zonyl FSA®. Next, 7.5 mL of acetic acid/sodium acetate buffer solution (pH = 4) were added to bring the total mixture volume up to 10 mL. The mixture was then centrifuged for 10 min at 2000 rpm to separate the sedimented (Zonyl FSA®) phase and the supernatant (aqueous) phase. Each rendered post-centrifugal phase volume was $V_{aq.}$ = 9.95 mL and VZonyl FSA® = 50 µL. The concentration of elemental ions in the aqueous phase was determined by ICP-OES.

The extraction percentage (E %) of elemental ions was calculated as

$$E\,\% = 100\,(1 - C_{aq.}/C_{Total}) \tag{1}$$

where C_{Total} and $C_{aq.}$ represent the total concentration of the added element and the concentration of the element in the aqueous phase after separation, respectively. Of note, the error range of each element's extraction percentage is ±5% or less.

2.3.2. The Recovery of Elements from the Zonyl FSA® Phase and the Redissolution of the Zonyl FSA® Phase

A water-immiscible Zonyl FSA® phase has retained the state of a stable liquid phase in the aqueous solution, and the elemental ions were recovered by filtration. In the recovery of targets, rare earth elements (Sc, Y, and lanthanoids) were well extracted by the experiment described in Section 2.3.1, and they were thus selected as targets. A Zonyl FSA® phase (approximately 50 µL) was produced after rare earth elemental extraction, which held on the hydrophilic PTFE filter. Water-immiscible Zonyl FSA® was filtered by adding 1 mL of sodium hydroxide and THF solution, which were mixed at a volume ratio of 1:1. The elements were trapped on the filter as hydroxides, the recovery percentage of elemental ions in the filtrate phase was determined by ICP-OES. Furthermore, the rare earth elements were recycled again by the same procedure outlined in Section 2.3.1 using the filtrated Zonyl FSA®.

3. RESULTS AND DISCUSSION

3.1. The Extraction of THF within the Phase Separation Phenomenon

As solubilizing agents for Zonyl FSA®, acetone and THF are examined. THF was selected from the viewpoint of its concentration time and operability, because it was already reported that the sedimented volume and its associated viscosity increase in the case of using acetone [22]. The relationship of the volume percentage of THF in the sedimented volume showed linear behavior in the range of 10–35 vol. % for the THF volumes shown in Figure 2 below. In the case of below 10 vol. % THF, the Zonyl FSA® phase precipitated fine solid particles. Moreover, Zonyl FSA® did not form a sedimented phase in the case of THF exceeding 50 vol. %. Therefore, THF volume in this system was ultimately deemed appropriate for 10 vol. %.

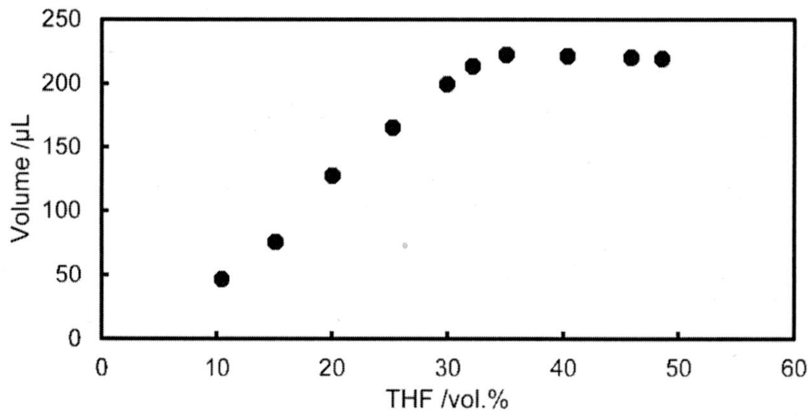

Figure 2. Relationship between the volume percentage of THF and the volume of sedimented phase. (Zonyl FSA®) = 10 vol. %; pH = 4.

3.2. Screening of Elements

The experimental procedure was performed according to Section 2.3.1. The percentage of extracted element ions in the sedimented phase is shown in Figure 3. Here, some element ions in the sample solution are estimated to exist as different type species such as a free metal ion, an anion complex such as gold chloride complex or platinum chloride complex, and an oxo anion species such as boric acid or phosphoric acid. Gallium, zirconium, palladium, silver, platinum, and rare earth elements (except for praseodymium) were obtained with a high extraction percentage; however, gold, indium, and germanium were obtained with a very low extraction percentage. With respect to rare earth elements, most rare earth elements are stable in the trivalent state, and they do not produce a hydroxide unless the pH range is of weak basicity to neutrality, e.g., since the solubility product constant (K_{sp}) of lanthanum is 1.0×10^{-19} [27], it does not precipitate as a hydroxide during the extraction procedure (pH = 4.0). On the other hand, only cerium was obtained with a notably low extraction percentage. This is because cerium is thought to exist in trivalent and tetravalent states in an aqueous solution with a pH of approximately 4. Based on this reason, tetravalent cerium is precipitated as a hydroxide in an acidic condition and, thus, is still susceptible to hydrolysis [28,29]. Therefore, the extraction into the sedimented Zonyl FSA® phase can present inherent challenges because of the fact that tetravalent cerium has usually already hydrolyzed at the time of extraction in a pH equal to 4.

3.3. Stripping of Rare Earth Elements

The extraction behavior of elements into the Zonyl FSA® liquid phase was examined by focusing on rare earth elements. Filtration operation by the filter was evaluated for the sedimented phase containing rare earth elements. The sedimented phase was held/maintained by a microsyringe on a Teflon filter (Pore size: 1 μm). The rare earth elements were collected on the filters as hydroxides by a 1:1 mixed solvent of 1 mL THF and 1 M sodium hydroxide aqueous solution, whereas the Zonyl FSA® was dissolved in the procedure. That rare earth elements had been stripped was confirmed by measuring Zonyl FSA® aqueous solution that was dissolved and filtrated using ICP-OES (Table 1).

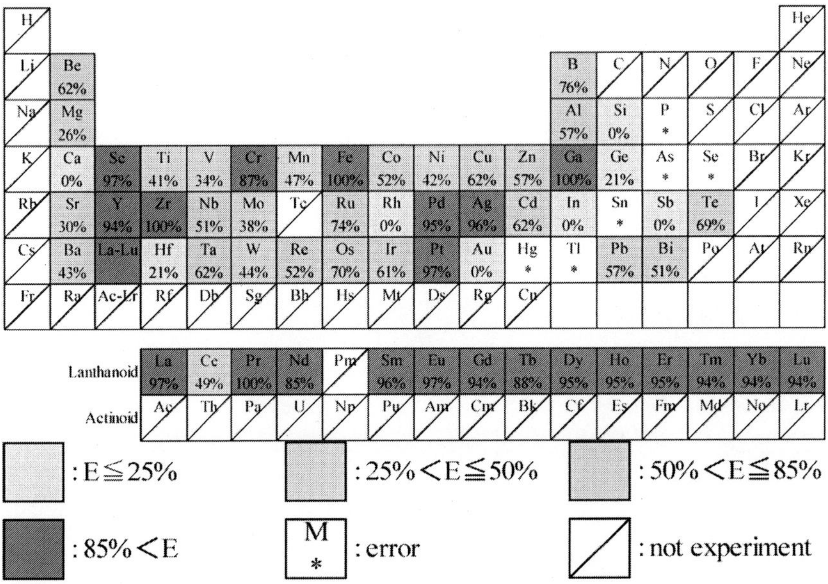

Figure 3. Summary of the extraction percentages for elemental ions with Zonyl FSA®.

Table 1. Stripping percentages of rare earth elements from the Zonyl FSA® phase.

Metal	Sc	Y	La	Ce	Pr	Nd	Sm	Eu	Gd	Tb	Dy	Ho	Er	Tm	Yb	Lu
Recovery %	81	92	88	90	91	88	91	92	92	91	92	92	84	92	92	92

3.4. Reuse of Spent Zonyl FSA®

The subject rare earth elements could be separated from the sedimented Zonyl FSA® phase by filtration, and the sedimented liquid phase similar to what existed before extraction was formed by adding a buffer solution (pH = 4) in the filtrate containing the Zonyl FSA®. Based on this phenomenon, the stripping experiments for rare earth elements were performed again using the spent Zonyl FSA®. The related findings are shown below in Table 2. Part of the heavy elements has a higher extraction percentage in the second extraction compared to the first extraction. Because part of the Zonyl FSA was lost during the recycling procedure, the component balance was changed in each fluorinated carbon chain number (6–8) of Zonyl FSA®. Thereby, the extraction selectivity for heavy elements was relatively increased. Zonyl FSA® was found to be reusable for the extraction and stripping of rare earth elements.

Table 2. Comparison of extraction percentages in first and second extraction times.

Metal	Sc	Y	La	Ce	Pr	Nd	Sm	Eu	Gd	Tb	Dy	Ho	Er	Tm	Yb	Lu
1st Extraction %	97	94	97	49	100	85	96	97	94	88	95	95	95	94	94	94
2nd Extraction %	80	100	93	53	100	84	100	100	100	94	100	100	94	100	100	100

3.5. Overview of Separation and Recycling System of Rare Earth Elements Using Zonyl FSA®

Extracting rare earth elements by Zonyl FSA® is a plausible recovery technique using a simple filtration operation, and the spent Zonyl FSA® can be reused. Therefore, a separation and recycling system for rare earth elements using Zonyl FSA®was constructed, with the subject scheme shown below in Figure 4. In this system, the Zonyl FSA® and THF were added to metal solutions containing rare earth elements (depicted in the figure), with the solution resultantly forming a homogeneous state by gentle shaking. In addition, a weak acid buffer solution is introduced into the solution to extract the rare earth elements, and the Zonyl FSA® is subsequently aggregated by centrifugation for a period. The sedimented Zonyl FSA® phase is ultimately retained on the filter. Then, only Zonyl FSA® is filtrated and the rare earth elements are recovered on the filter as hydroxides. In contrast, Zonyl FSA® solution is obtained as a filtrate to around a neutral pH (pH ≥ 6.5). Accordingly, it is capable as a second recycling product to be used again for separation and recovery, as demonstrated below in the figure.

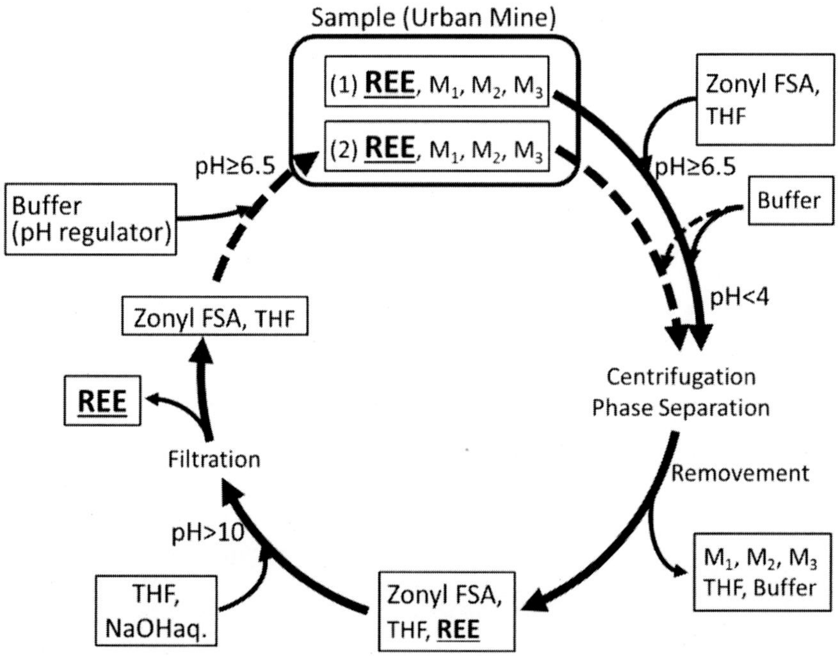

Figure 4. Separation and recycling system for rare earth elements using Zonyl FSA®. REE: rare earth elements; M_x: coexistent metals.

4. CONCLUSIONS

A homogeneous liquid-liquid extraction method (HoLLE) for metal elements using a Zonyl FSA® fluorosurfactant instead of a chelating agent was proposed. As a result, metal elements such as palladium, platinum, and rare earth elements were extracted at a very high percentage. This method is highly capable of extracting and stripping (with high efficiency) rare earth elements by HoLLE via the use of Zonyl FSA®. This method is expected to be a preprocessing technique for separating and recycling rare earth elements in electrical appliance components. Moreover, the developed process is capable of both the recycling of rare earth elements and the reuse of Zonyl FSA® as an extraction agent.

ACKNOWLEDGMENTS

The authors thank Toshinori Fukazawa for assistance with the screening experiment of metals. This work was supported by JSPS KAKENHI Grant Numbers 15H02847. The authors would like to thank Enago (www.enago.jp) for the English language review.

AUTHOR CONTRIBUTIONS

S.S. and S.I. conceived the experiments, and wrote the paper; S.S. designed experiments with contributions from S.I., O.O., T.K. and H.Y.; H.Y. was measured of metals by ICP-OES; S.S. performed experiments and interpretation.

CONFLICTS OF INTEREST

The authors declare no conflict of interest.

REFERENCES

1. Yang, J.; Kubota, F.; Baba, Y.; Kamiya, N.; Goto, M. One step effective separation of platinum and palladium in an acidic chloride solution by using undiluted ionic liquids. *Solvent Extr. Res. Dev. Jpn.* 2014, *21*, 129–135.
2. Yang, J.; Kubota, F.; Baba, Y.; Kamiya, N.; Goto, M. Separation of precious metals by using undiluted ionic liquids.*Solvent Extr. Res. Dev. Jpn.* **2014**, *21*, 89–94.
3. Saitoh, T.; Sugiura, Y.; Asano, K.; Hiraide, M. Chitosan-conjugated thermos-responsive polymer for the rapid removal of phenol in water. *React. Funct. Polym.* **2009**, *69*, 792–796.
4. Kato, T.; Igarashi, S.; Ohno, O.; Watanabe, Y.; Murakami, K.; Takemori, T.; Yamaguchi, H.; Ando, R. Separation and recovery properties of rare earth elements using a pH-

sensitive polymer having benzoic acid substituent group.*Bunseki Kagaku* **2012**, *61*, 235–242.

5. Igarashi, S.; Saito, S.; Kato, T.; Okano, G.; Yamaguchi, H. Development of separation and recovery system of gold and rare earth elements using stimuli-responsive polymers—G-MOVE system and La-VEBA system. *J. Surf. Finish. Soc. Jpn.* **2012**, *63*, 630–632.

6. Saito, S.; Igarashi, S.; Yamaguchi, H. Selective collection characteristics and separation/recovery method of gold(III), silver(I) and platinum(II) with pH-sensitive polymer under addition of L-ascorbic acid. *Bunseki Kagaku* **2014**, *63*, 791–795.

7. Inoue, K.; Ohto, K.; Yoshizuka, K.; Shinbaru, R.; Baba, Y.; Kina, K. Adsorption behavior of metal ions on some carboxymethylated chitosans. *Bunseki Kagaku* **1993**, *42*, 725–731.

8. Parajuli, D.; Kawakita, H.; Inoue, K.; Ohto, K.; Kajiyam, K. Persimmon peel gel for the selective recovery of gold.*Hydrometallurgy* **2007**, *87*, 133–139.

9. Gurung, M.; Adhikari, B.B.; Kawakita, H.; Ohto, K.; Inoue, K. Recovery of Au(III) by using low cost adsorbent prepared from persimmon tannin extract. *Chem. Eng. J.* **2011**, *174*, 556–563.

10. Igarashi, S.; Yotsuyanagi, T. A novel homogeneous liquid-liquid Extraction by pH dependent phase separation with fluorocarbon ionic surfactant. In Proceedings of the Symposium on Solvent Extraction 1988, Tokyo, Japan, 5–7 December 1988; pp. 175–180.

11. Igarashi, S.; Yotsuyanagi, T. Temperature dependent phase transformation in the homogeneous liquid-liquid extraction with fluorocarbon surfactant. In Proceedings of the Symposium on Solvent Extraction 1989, Sendai, Japan, 9–11 November 1989; pp. 51–54.

12. Igarashi, S.; Yotsuyanagi, T. New homogeneous liquid-liquid extraction by phase separation and transformation with fluorocarbon surfactant and quaternary ammonium salt. In Solvent extraction 1990, Proceedings of the

International Solvent Extraction Conference (ISEC '90), Kyoto, Japan, 16–21 July 1992; Sekine, T., Ed.; Elsevier: New York, NY, USA, 1992; pp. 1725–1730.

13. Wilkes, J.S.; Zaworotko, M.J. Air and water stable 1-ethyl-3-methylimidazolium based ionic liquids. *J. Chem. Soc. Chem. Commun.* **1992**, *13*, 965–967.

14. Yamaguchi, H. Did the study of ionic liquids begin from analytical chemistry? *Bunseki* **2007**, *11*, 608–609.

15. Igarashi, S.; Yotsuyanagi, T. Homogeneous liquid-liquid extraction by pH dependent phase separation with a fluorocarbon ionic surfactant and its application to the preconcentration of porphyrin compounds. *Mikrochimica Acta* **1992**, *106*, 37–44.

16. Oshite, S.; Igarashi, S. Homogeneous liquid-liquid extraction using perfluorooctanesulfonic acid and calcium and its application to the separation and recovery vitamin B_{12}. *J. Chem. Technol. Biotechnol.* **1999**, *74*, 1183–1187.

17. Oshite, S.; Furukawa, M.; Igarashi, S. Homogeneous liquid-liquid extraction method for the selective spectrofluorimetric determination of trace amoutns of tryptophan. *Analyst* **2001**, *126*, 703–706.

18. Takahashi, A.; Ueki, Y.; Igarashi, S. Highly efficient homogeneous liquid-liquid extraction of rare earth metal ions from perfluorocarboxylate surfactant solutions using ion-pair phase separation systems. *Solvent Extr. Res. Dev. Jpn.* **2001**, *8*, 235–240.

19. Hoogerstraete, T.V.; Ohghena, B.; Binnemans, K. Homogeneous liquid-liquid extraction of rare earths with the betaine-batainium bis(trifluoromethylsulfonyl)imide ionic liquid system. *Int. J. Mol. Sci.* **2013**, *14*, 21353–21377.

20. Takagai, Y.; Igarashi, S. Homogeneous liquid-liquid extraction method as simple and powerful preconcentration for capillary gas chromatography and capillary electrophoresis. *Am. Lab.* **2002**, *34*, 29–30.

21. Fuchimukai, J.; Yamaguchi, H.; Meguro, Y.; Kubota, T.; Igarashi, S. Highly efficient homogeneous liquid-liquid

extraction of lanthanoid ions in a strong acidic solution. *Solvent Extr. Res. Dev. Jpn.* **2006**, *13*, 139–146.

22. Sudo, T.; Igarashi, S. Homogeneous liquid-liquid extraction method for spectrofluorimetric determination of chlorophyll a. *Talanta* **1996**, *43*, 233–237.

23. Ghiasvand, A.R.; Moradi, F.; Sharghi, H.; Hasaninejad, A.R. Determination of silver(I) by electrothermal-AAS in a microdroplet formed from a homogeneous liquid-liquid extraction system using tetraspirocyclohexylcalix[4]pyrroles.*Anal. Sci.* **2005**, *21*, 387–390.

24. Ghiasvand, A.R.; Shadabi, S.; Kakanejadifard, A.; Khajehkoolaki, A. Synthesis of a new α-dioxime derivative and its application for selective homogeneous liquid-liquid extraction of Cu(II) into a microdroplet followed by direct GFAAS determination. *Bull. Korean Chem. Soc.* **2005**, *26*, 781–785.

25. Kato, T.; Igarashi, S.; Ishiwatari, Y.; Furukawa, M.; Yamaguchi, H. Separation and concentration of indium from a liquid crystal display via homogeneous liquid-liquid extraction. *Hydrometallurgy* **2013**, *137*, 148–155.

26. Yamaguchi, H.; Itoh, S.; Igarashi, S.; Kobayashi, T. Homogeneous liquid-liquid extraction of metal-1,10-phenanthoroline chelates in a weak acidic solution. *Bunseki Kagaku* **2005**, *54*, 227–230.

27. Sinha, S.P. *Complexes of the Rare Earths*; Pergamon Press Ltd.: Oxford, UK, 1966; p. 26.

28. Suzuki, Y. *Kidorui no Hanashi [The Story of Rare Earth Elements]*; Shokabo Publishing Co., Ltd.: Chiyoda, Japan, 1998; pp. 66–72.

29. Burgess, J. *Metal Ions in Solution*; Ellis Horwood Ltd.: New York, NY, USA, 1978; pp. 266–267.

13

CHAPTER

Use of Phosphate Solubilizing Bacteria to Leach Rare Earth Elements from Monazite-Bearing Ore

Doyun Shin [1,2,*], Jiwoong Kim [1], Byung-su Kim [1,2], Jinki Jeong [1,2] and Jae-chun Lee [1,2]

[1]*Mineral Resources Resource Division, Korea Institute of Geoscience and Mineral Resources (KIGAM), Gwahangno 124, Yuseong-gu, Daejeon 305-350, Korea*

[2]*Department of Resource Recycling Engineering, Korea University of Science and Technology, Gajeongno 217, Yuseong-gu, Daejeon 305-350, Korea*

ABSTRACT

In the present study, the feasibility to use phosphate solubilizing bacteria (PSB) to develop a biological leaching process of rare earth elements (REE) from monazite-bearing ore was determined. To predict the REE leaching capacity of bacteria, the phosphate solubilizing abilities of 10 species of PSB were determined by halo zone formation on Reyes minimal agar media supplemented with bromo cresol green together with a phosphate solubilization test in Reyes minimal liquid media as the screening studies. Calcium phosphate was used as a model mineral phosphate. Among the test PSB strains, *Pseudomonas fluorescens*, *P. putida*, *P. rhizosphaerae*, *Mesorhizobium ciceri*, *Bacillus megaterium*, and *Acetobacter aceti* formed halo zones, with the zone of *A.*

aceti being the widest. In the phosphate solubilization test in liquid media, *Azospirillum lipoferum*, *P. rhizosphaerae*, *B. megaterium*, and *A. aceti* caused the leaching of 6.4%, 6.9%, 7.5%, and 32.5% of calcium, respectively. When PSB were used to leach REE from monazite-bearing ore, ~5.7 mg/L of cerium (0.13% of leaching efficiency) and ~2.8 mg/L of lanthanum (0.11%) were leached by *A. aceti*, and *Azospirillum brasilense*, *A. lipoferum*, *P. rhizosphaerae* and *M. ciceri* leached 0.5–1 mg/L of both cerium and lanthanum (0.005%–0.01%), as measured by concentrations in the leaching liquor. These results indicate that determination of halo zone formation was found as a useful method to select high-capacity bacteria in REE leaching. However, as the leaching efficiency determined in our experiments was low, even in the presence of *A. aceti*, further studies are now underway to enhance leaching efficiency by selecting other microorganisms based on halo zone formation.

Keywords: bioleaching; monazite; phosphate solubilizing bacteria; rare earth element

1. INTRODUCTION

Rare earth elements (REE) have been increasingly used in the fields of optics, permanent magnetism, electronics, superconductor technology, hydrogen storage, medicine, nuclear technology, secondary battery technology, and catalysis [1,2,3]. The minerals monazite (a phosphate mineral) and bastnasite (a fluorocarbonate mineral) are the main sources of REE in nature. Generally, monazite contains ~70% rare earth metal oxide, with the rare earth fraction comprising 20%–30% Ce_2O_3, 10%–40% La_2O_3, and substantial amounts of neodymium, praseodymium, and samarium. The thorium content is in the range of 4%–12% [2,3].

Caustic soda decomposition and concentrated sulfuric acid digestion have been widely used to decompose monazite for many decades [2,4]. Due to its high chemical and thermal stability, monazite is very difficult to decompose; therefore, it is essential to eliminate the phosphate present in the ore by chemically attacking the mineral with sulfuric acid or sodium hydroxide at high temperature, so as to enhance the capacity to dissolve the REE. The sulfuric acid process results in a loss of phosphate as H_3PO_4, corrosion of the processing facilities, toxic gas and wastewater generation, as well as yielding impure products; the process is therefore no longer in commercial use. Caustic soda decomposition has some advantages in terms of the recovery of unreacted alkali and phosphorous, low energy consumption, and simplicity; however, the process also has limitations, such as the need for high-grade ore sources [5,6].

Biohydrometallurgical technology is an attractive alternative emerging green technology for the recovery of metals due to its environmental friendly, simple, and economic processing. However, very few works have been published on the biological recovery of rare earth metals, in particular, from monazite. Recently thorium, uranium, and REE extraction by microorganisms from monazite concentrate was reported [7,8]. The authors used *Aspergillus ficuum*, organic acid producing fungi, and *Pseudomonas aeruginosa*, organic acid/siderophore producing bacteria. They found that those microorganisms produced citric, oxalic, or 2-ketogluconic acid and dissolved 55% and 47% of REE from monazite by *A. ficuum* and *P. aerunoginosa*, respectively. Another study on REE leaching from phosphate minerals apatite and monazite by organic acids such as citric, oxalic, phthalic, and salicylic was also published, even though chemical organic acids were used in this study, which were not biologically produced [9]. Organic acid producing microorganisms secrete organic acids such as malic, gluconic, or oxalic acids [10,11], and the mechanism of metal dissolution by the microorganisms is both of acidolysis (protons dissociated from the organic acids) and complexolysis (metal-complexing anions from the acids) [12].

In a soil improvement process, phosphate solubilizing bacteria (PSB) have been widely used because they affect the release of phosphorous (P) from inorganic and organic P pools through solubilization and mineralization [13,14,15] by organic acid production. In the present study, we investigated the feasibility to use PSB in REE leaching from monazite-bearing ore as an exploratory study. Screening studies were performed by determining halo zone formation on agar media and phosphate solubilization in liquid media to predict REE leaching capacity of bacteria. The REE leaching abilities of the PSB from monazite-bearing ore were determined and compared with the screening data.

2. EXPERIMENTAL SECTION

2.1. Bacterial Strains and Culture Conditions

Ten PSB strains, *Pseudomonas rhizosphaerae* DSM 16299^T [16], *P. putida* DSM 291^T, *P. fluorescens* DSM 50090^T [17], *Bacillus megaterium* DSM 32^T [18], *Paenibacillus polymyxa* DSM 36^T [14], *Ensifer meliloti* DSM 30135 [19], *Azospirillum brasilense* DSM 1690^T, *A. lipoferum* DSM 1842 [17], *Mesorhizobium ciceri*, and *Acetobacter aceti* DSM 2002 were selected, purchased from Deutsche Sammlung von Mikroorganismen und Zellkulturen (DSMZ; Braunschweig, Germany), and maintained on nutrient agar (NA) at 30 °C. For the organic acid production test, the phosphate solubilization test, and the RE leaching study, Reyes minimal medium (in 1 L: 0.4 g NH_4Cl; 0.78 g KNO_3; 0.1 g NaCl; 0.5 g $MgSO_4 \cdot 7H_2O$; 0.1 g $CaCl_2 \cdot 2H_2O$; 0.5 mg $FeSO_4 \cdot 7H_2O$; 1.56 mg $MnSO_4 \cdot H_2O$; 1.40 mg $ZnSO_4 \cdot 7H_2O$) [20] was used. Glucose or sucrose was added at 30 g/L as a carbon source and $Ca_3(PO_4)_2$ (5.39 g/L), alkaline metal phosphate, was used as a sole phosphate source. The viable bacterial cell number was measured by the plate counting method on NA at 30 °C.

2.2. Comparison of Halo Zone Formation, Organic Acid Production, and Phosphate Solubilization by Phosphate Solubilizing Bacteria

Halo zone formation by test PSB strains were tested using Reyes minimal medium supplemented with 1.5% agar, 5.39 g/L $Ca_3(PO_4)_2$ as a mineral phosphate, 30 g/L glucose or sucrose as a carbon source, and bromocresol green (BCG) as a pH indicator. The pH of the medium was adjusted to 6.5 using 1 N KOH. A 0.5% BCG solution was prepared in 70% ethanol as a stock solution and a 0.5 mL aliquot of the solution was added to 100 mL of Reyes minimal agar medium before autoclaving. Twenty microliters of PSB pregrown in nutrient broth was dot inoculated onto the surface of the agar plate prepared as above, incubated at 30 °C for 48 h. The total halo diameter and colony diameter were measured; halo size was calculated by subtracting the colony diameter from the total halo diameter.

A quantitative estimate of organic acid production by the PSB was performed using a 50-mL conical tube containing 30 mL of Reyes minimal medium supplemented with 0.45 g/L KH_2PO_4 and 30 g/L glucose, and inoculated with the bacterial strain (3 mL inoculum with approximately 10^8 CFU (colony forming unit)/mL). To measure the organic acid producing ability of the test PSB without any interference, KH_2PO_4 was used as an easily edible form of phosphate source because the availability of a phosphate source may affect organic acid production. An autoclaved uninoculated medium served as a control. The vials were incubated for 11 days in a shaking incubator at 30 °C and 180 rpm. At 1-day intervals, individual cultures were centrifuged at 5000 g for 30 min, and the supernatant was filtered with a 0.22-μm pore syringe filter. Concentrations of citric acid ($C_6H_8O_7$), malic acid ($C_4H_6O_5$), tartaric acid ($C_4H_6O_6$), and acetic acid ($C_2H_4O_2$) in the culture were determined by high-performance liquid chromatography (HPLC) (2690 Separation Module, Waters Alliance System, Milford, MA, USA) with 20 mM NaH_2PO_4 (pH 2.7, adjusted with 20% phosphoric acid), using an Atlantis T3 column (5 μm; 4.6 mm × 250 mm), and quantified using a Waters 996 Photo-diode Array Detector at 210 nm.

Phosphate solubilization by the test PSB in liquid media was determined in Reyes minimal medium supplemented with $Ca_3(PO_4)_2$ (5.39 g/L) and glucose (30 g/L). The cultivation procedure was as described above, except for the incubation time (*i.e.*, 5 days). Total calcium concentrations were determined by inductively coupled plasma atomic emission spectrometry (ICP-AES) (Optima 5300DV, Jobin Yvon JY 38, ICAP 6500 DVO; Perkin Elmer, Shelton, CT, USA); phosphate concentrations were determined by ion chromatography (IC) (ICS-3000, Dionex, Sunnyvale, CA, USA). Values reported are the average of three replicates.

2.3. Ore Characterization

The elemental composition of the raw monazite-bearing ore used in this study is shown in Table 1, which was measured by alkali fusion with Na_2CO_3 for silica, permanganametric titration for calcium, and ICP after multi-acid digestion for the other elements [21]. The mineralogical and morphological data of the ore were obtained by X-Ray Diffraction (XRD; Model RTP 300 RC, Rigaku Co., Tokyo, Japan) and a scanning electron microscope (SEM, JEOL 6400; JEOL Ltd., Tokyo, Japan) equipped with energy dispersive spectroscopy (EDS). The semi-quantification spot analysis SEM-EDS was performed using Thin Film Standardless Standard Quantitative Analysis (Oxide).

Table 1. Elemental analysis of monazite-bearing ore used in this study.

Element	Ce	La	Nd	Pr	SiO_2	Al_2O_3	Fe_2O_3	CaO	MgO	K_2O	Sr	TiO_2	P_2O_5
Content (wt%)	3.50	2.09	0.75	0.21	12.77	0.73	39.80	7.52	6.26	<0.001	0.47	0.02	5.33

2.4. Monazite-Bearing Ore Leaching by Phosphate Solubilizing Bacteria

The monazite-bearing ore was ground and then sieved below 150 µm size. A solution of 30 mL Reyes minimal medium (prepared as above) containing 30 g/L of glucose was added to 50-mL conical tubes containing 5 g of the autoclaved ore. Leaching was run for 9 days in a shaking incubator at 30 °C and 180 rpm. On each of the 9 days, samples were taken and prepared independently for further analysis. Total cerium and lanthanum concentrations in the residue and the liquor were determined by ICP and the mineralogical change of the residue was analyzed by SEM-EDS and XRD.

3. RESULTS AND DISCUSSION

3.1. Comparison of Phosphate Solubilization by Phosphate Solubilizing Bacteria

Ten PSB strains (as stated above) were selected based on the literature reviews of their potential for phosphate solubilization. The halo zone formation was determined as a screening study to investigate the phosphate solubilizing ability of the PSB. Due to the production of organic acids in the surrounding medium, the halo zone on the agar plate was formed by converting insoluble phosphate into soluble phosphate [22]. The converting efficiencies vary for different forms of phosphate (*i.e.*, $Ca_3(PO_4)_2$, $AlPO_4$, and $FePO_4$); among them, calcium phosphate was most aggressively attacked by PSB. Higher phosphate solubilization from calcium phosphate than from aluminum and iron phosphate minerals has been reported, based on the solubility equilibrium and acidity constants of these compounds [23], thus $Ca_3(PO_4)_2$ was used in this study as a mineral phosphate. Bromocresol green (BCG) was also added in Reyes minimal medium supplemented with glucose and sucrose to improve the clarity and visibility of the yellow-colored halo zone [11]. The test PSB formed different sizes of distinct halo zones on the media depending on the bacterial strain and carbon source (Table 2). The widest halo zone was observed in the presence of *A. aceti* (diameter, 41.0 mm) on media supplemented with glucose; *P. rhizosphaerae*, *P. fluorescens*, *B. megaterium*, and *M. ciceri* also formed halo zones, while *P. polymyxa*, *E. meliloti*, *A. brasilense*, and *A. lipoferum* did not form halo zones on the media supplemented with either glucose or sucrose. Wider halo zones were observed in most samples when glucose was present rather than when sucrose was present; therefore, glucose was used as a carbon source throughout this study. Also, the use of modified Reyes agar medium supplemented with BCG and glucose was the most effective condition to determine halo zone formation.

Table 2. Comparison of tricalcium phosphate solubilization by phosphate solubilizing bacteria (PSB) in Reyes minimal agar medium supplemented with bromocresol green (BCG) and 30 g/L of glucose or sucrose, associated with microbial growth for 24 h.

Bacterial Species	Halo Size (mm) *	
	Glucose	Sucrose
P. rhizosphaerae DSM 16299^T	13.5	13.4
P. putida DSM 291^T	-**	7.2
P. fluorescens DSM 50090$_T$	5.6	7.8
B. megaterium DSM 32^T	4.7	12.1
P. polymyxa DSM 36^T	-	-
E. meliloti DSM 30135^T	-	-
A. brasilense DSM 1690^T	-	-
A. lipoferum DSM 1842	-	-
M. ciceri	5.1	4.4
A. aceti DSM 2002	41.0	12.58

* The total halo diameter and colony diameter were measured; halo size was calculated by subtracting the colony diameter from the total halo diameter. ** "-" represents no halo zone formation.

Concentrations of solubilized phosphate and calcium ion in the Reyes minimal medium supplemented with glucose and $Ca_3(PO_4)_2$ were analyzed for a period of five days, to compare the phosphate solubilizing ability of PSB. The viable PSB cell numbers were maintained at approximately 10^7 CFU/mL medium during the experimental periods. The leaching efficiency was calculated by subtracting the mass of calcium and phosphate in the uninoculated control from the mass in the leaching liquor and dividing the leached mass by the total mass. The leaching efficiencies were shown to be highest generally within two to three days of incubation; thereafter, concentrations of leached phosphate and calcium decreased, possibly due to uptake by the PSB or absorption onto bacterial surfaces. The highest leaching efficiencies after three days of leaching were presented by *A. lipoferum*, *P. rhizosphaerae*, and *B. megaterium* (6.4%, 6.9%, and 7.5%, respectively, for calcium; and 3.7%, 6.8%, and 5.7%, respectively, for phosphate (Figure 1); leaching efficiencies by *A. brasilense* and *E. meliloti* were less than 2% for calcium; as expected, *A. aceti* showed the highest leaching efficiency of calcium (~32.5%).

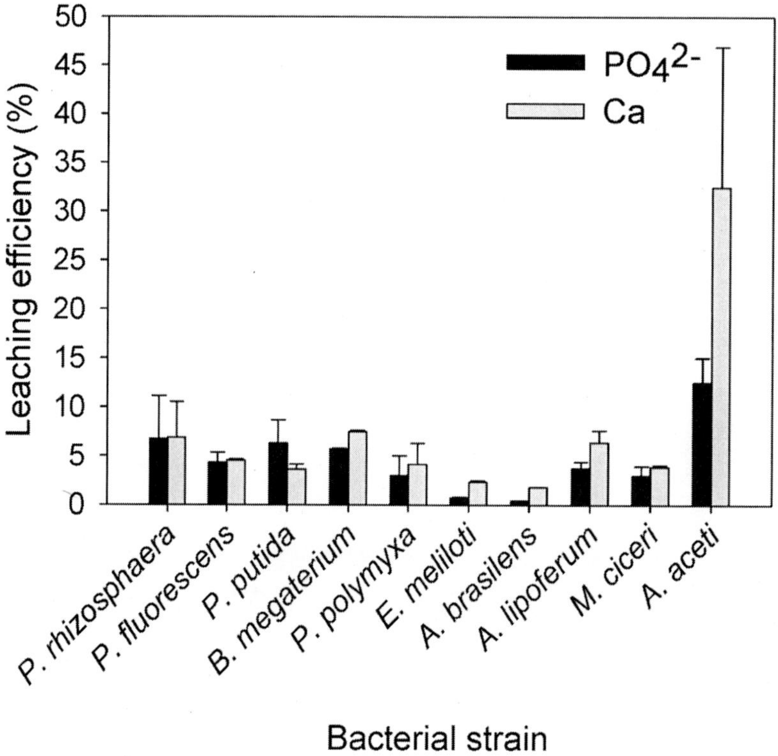

Figure 1. Phosphate and calcium solubilization from $Ca_3(PO_4)_2$ in Reyes basal liquid medium, associated with microbial growth after 3 days.

Inorganic phosphate solubilization by PSB depends on organic acid secretion from PSB [14]. Citric, malic, tartaric, and acetic acids were measured by HPLC in 30-mL of the PSB cultures containing KH_2PO_4 and glucose (Figure 2). The test PSB strains showed organic acid production below ~0.2 mmol, except for *A. aceti*. The maximum malic acid produced by *A. brasilense*, *A. lipoferum*, *M. ciceri*, and *P. rhizosphaerae* occurred at 0.07 mmol at six days, 0.09 mmol at nine days, 0.06 mmol at eight days, and 0.18 mmol at six days, respectively (maximum production). Acetic acid was produced by *A. aceti*, with concentrations reaching a maximum of 15.8 mM (0.48 mmol) at four days of incubation. Hwangbo *et al.* [24] reported that*Enterobacter intermedium* produced gluconic

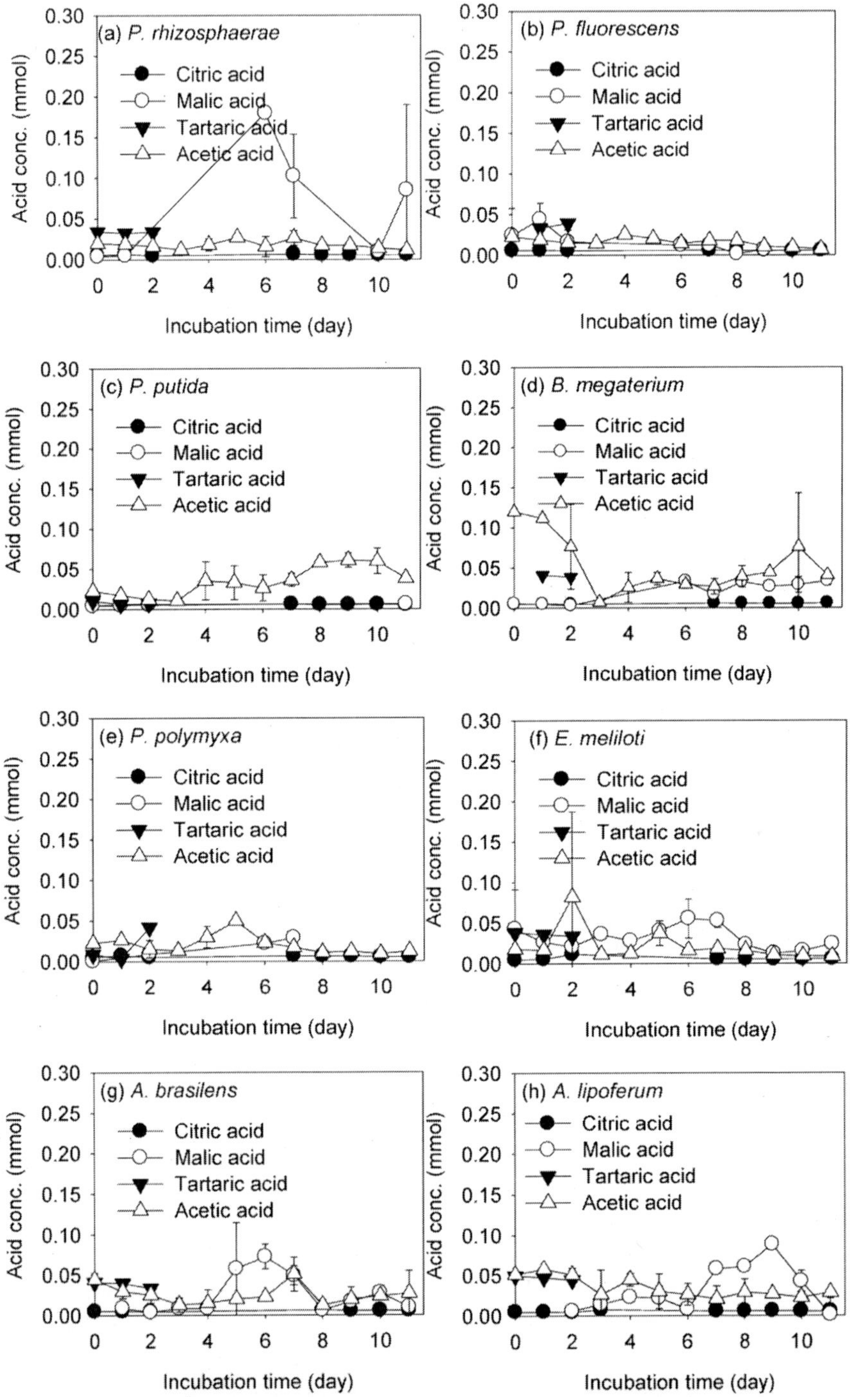

(a) P. rhizosphaerae
(b) P. fluorescens
(c) P. putida
(d) B. megaterium
(e) P. polymyxa
(f) E. meliloti
(g) A. brasilens
(h) A. lipoferum
Citric acid
Malic acid
Tartaric acid
Acetic acid
Acid conc. (mmol)
Incubation time (day)

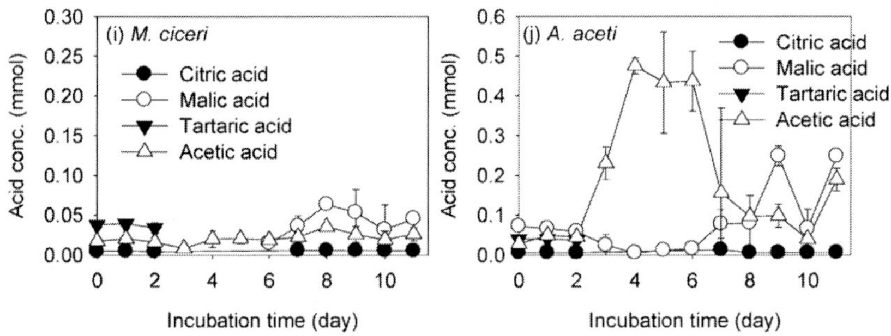

Figure 2. Organic acid production of (**a**) *P. rhizosphaerae* DSM 16299T, (**b**) *P. fluorescens*DSM 50090T, (**c**) *P. putida* DSM 291T, (**d**) *B. megaterium* DSM 32T, (**e**) *P. polymyxa* DSM 36T, (**f**) *E. meliloti* DSM 30135, (**g**) *A. brasilense* DSM 1690T, (**h**) *A. lipoferum* DSM 1842, (**i**)*M. ciceri*, and (**j**) *A. aceti* in Reyes basal liquid medium supplemented with K_2HPO_4, associated with microbial growth for 11 days.

acid from glucose and subsequently converted it to 2-ketogluconic acid, which has a strong ionic strength and therefore easily solubilizes rock phosphate into soluble forms. The same situation was reported when Acetobacter was used to produce gluconic acid [25]. Because of the conversion of gluconic acid as stated above, gluconic acid might not be detected in the microbial cultures in this study. After six days, the amount of malic and acetic acid decreased. The pH of the media during incubation showed similar trends with organic acid production (Figure 3) because the pH decrease is due to the excreted metabolites, which included proton from organic acids, amino acids, and other metabolites. The pH in the medium of *A. aceti* decreased to about pH 3 (corresponding pH to ~15 mM of acetic acid, which was the acid produced) until eight days of incubation, and subsequently increased. The acidities of the media for *A. brasilense*, *A. lipoferum*, and *P. rhizosphaerae* also showed similar trends with acetic acid, with pH decreasing (*i.e.*, 3–4) until six to seven days of incubation and increasing thereafter.

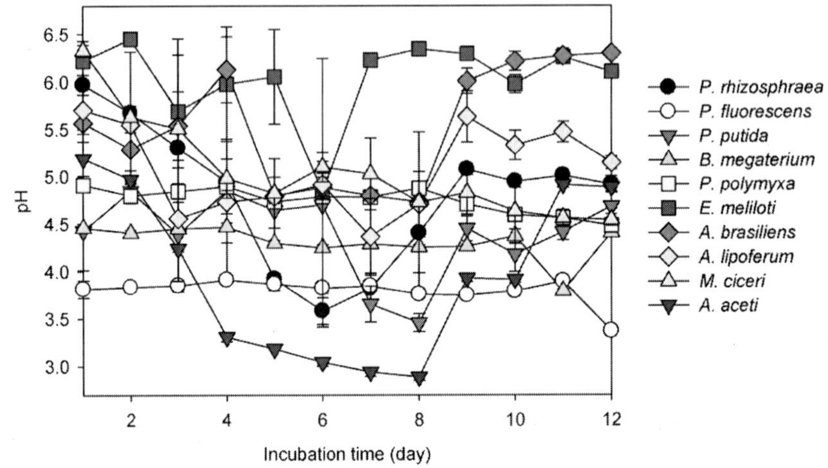

Figure 3. The pH changes induced by phosphate solubilizing bacteria (PSB) during incubation in Reyes minimal medium supplemented with K_2HPO_4.

3.2. REE Leaching by the Test PSB from the Monazite-Bearing Ore and Comparison with the Screening Study Data

Ten PSB were introduced to leach REE from monazite-bearing ore (Figure 4). We detected ~5.7 mg/L of cerium (0.13% of leaching efficiency) and 2.8 mg/L of lanthanum (0.11%) in the leaching liquor of *A. aceti*, and approximately 0.5–1 mg/L of cerium and lanthanum (0.005%–0.01%) in the leaching liquors of *A. brasilense*, *A. lipoferum*, *P. rhizosphaerae*, and *M. ciceri*. Since acetic and malic acid were produced by *A. aceti*, *A. brasilense*, *A. lipoferum*, *P. rhizosphaerae*, and *M. ciceri* as stated above, REE such as cerium and lanthanum may form complexes with these acids. When REE cations (Re_2O_3) are present in a system and acetic or malic acid is fully dissociated in aqueous solution, a complexation reaction may take place (at 25 °C) [26];

$$C_2H_4O_2 \leftrightarrow C_2H_3O_2^- + H^+ \ (pK_a = 4.756)$$

$$3C_2H_3O_2^- + Re_2O_3 \leftrightarrow Re(C_2H_3O_2)_3 \text{ (REE acetate complex)}$$

Also, since malic acid is a diprotic acid, the possible complexes of REE reaction with malate anions are as follows;

$C_4H_6O_5 \leftrightarrow C_4H_5O_5^- + H^+$ ($pK_a = 3.40$)

$3C_4H_5O_5^- + Re_2O_3 \leftrightarrow Re(C_4H_5O_5)_3$ (Rare earths malate complex)

$C_4H_5O_5^- \leftrightarrow C_4H_4O_5^{2-} + 2H^+$ ($pK_a = 5.11$)

$3C_4H_4O_5^{2-} + 2Re_2O_3 \leftrightarrow Re_2(C_4H_4O_5)_3$ (Rare earths malate complex)

The role of low molecular weight organic acids such as acetic, malic, oxalic, and citric acids in metal dissolution has been discussed in many studies [7,8,9,27,28]. The ligands of the organic acids play a dominant role in dissolving metals, and the differences in the effect of metal dissolution is related to their chemical structure. For instance, citric acid has great chelating ability with metals because citrate (*i.e.*, three carboxylic groups) can form stable chelates with 6-membered ring structures. While malic acid (dicarboxylic acid) is more stable than acetic acid (monocarboxylic acid) [29], REE leaching was observed in the *A. aceti* culture (where most of the leaching agent was acetic acid) because acetic acid production was higher than other organic acids production.

Mineralogical changes of the monazite-bearing ore before and after PSB leaching were also considered. As a variety of minerals that entrap monazite may be present, and these minerals may prevent contact of the organic acids produced by PSB as shown in the XRD patterns and the SEM images (Figure 5 and Figure 6; Table 3). The X-ray pattern of the monazite-bearing ore shows the presence of dolomite ($CaMg(CO_3)_2$), quartz (SiO_2), siderite ($FeCO_3$), pyrite (FeS_2), and monazite ((Ce, La, Pr, Nd, Th, Y)PO_4). The cerium and lanthanum-rich region (spot 1 and 2) had a spotty distribution surrounded by other minerals including Mg, Al, Si, or Fe (spot 3 and 4). High carbon ratio in the spots was because of epoxy impregnation in preparing flat-polished specimens. XRD patterns of the monazite-bearing ore before and after leaching in this study showed a reduction in all peak intensities (Figure 6). High peaks of carbonate minerals (dolomite ($CaMg(CO_3)_2$) and siderite ($FeCO_3$)) were found and these minerals might consume the organic acids produced by PSB and prevent REE dissolution.

Differently from the phosphate solubilization test in liquid media (Figure 1), *P. fluorescens*, *P. putida*, *B. megaterium*, and *P.*

polymyxa did not leach REE from the monazite-bearing ore. In the case of *A. brasilense* and *A. lipoferum*, even though they leached REE from the ore in spite of small leaching efficiency (Figure 4) and produced similar amounts of malic acid with *M. ciceri* and *P. rhizosphaerae* (Figure 2), they did not form halo zones (Table 2). This discrepancy sometimes happens because of the difference between cultivation in liquid media and on agar media. A liquid medium is preferred for the production of such organic acids because excreted products are readily available from a liquid culture and the cells are uniformly exposed to conditions of the medium [30]. However, the strongest leaching of REE was observed in the presence of *A. aceti*, as expected by both the results of the halo zone formation and the phosphate solubilization test. In the case of low-capacity bacteria, some errors may occur as stated above in determination of halo zone formation, but this method is still useful to roughly estimate phosphate solubilizing and organic acid producing abilities, as well as REE leaching capacities. Thus we believe that determination of halo zone formation can be used as a simple and rapid screening method to select high-capacity bacteria in REE leaching.

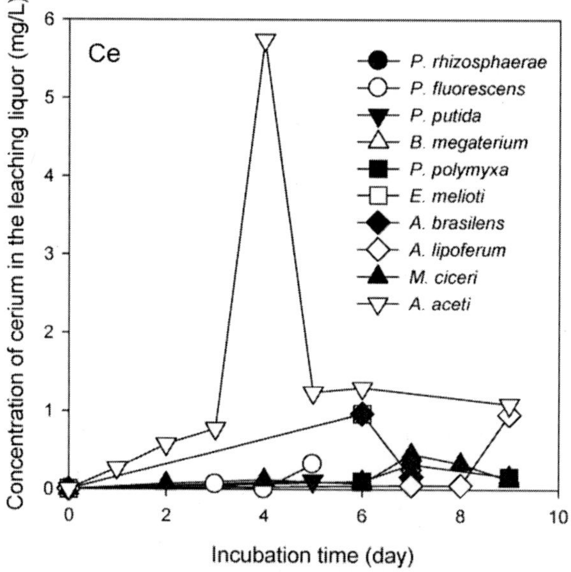

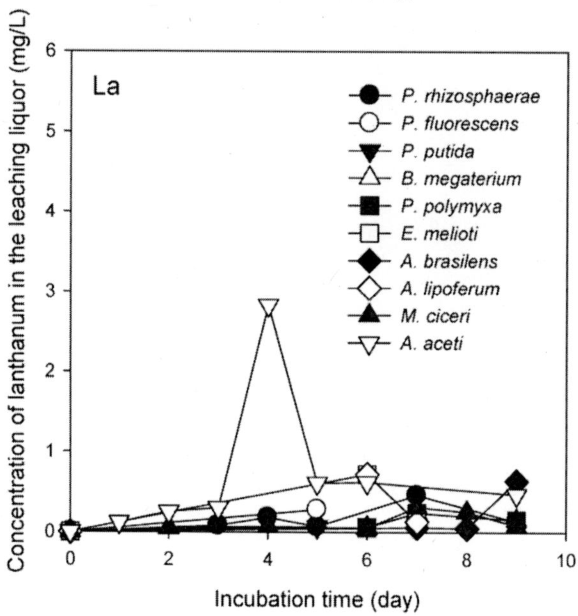

Figure 4. The concentrations of leached cerium and lanthanum from monazite-bearing ore by phosphate solubilizing bacteria (PSB) in Reyes basal liquid medium.

Table 3. Elemental compositions of the spots in Figure 5 by scanning electron microscopy/energy dispersive X-ray spectroscopy (SEM-EDS).

Spot No. in Figure 5	Mass %											
	C	O	Mg	Al	Si	P	S	Ca	Fe	La	Ce	Pr
1	82.1	4.88	1.67	-	-	-	-	7.06	-	0.18	4.1	-
2	46.0	22.7	1.28	-	5.4	-	8.2	-	-	3.36	12.5	0.62
3	61.3	16.4	1.05	0.24	8.77	2.15	-	-	9.26	-	-	-
4	70.6	15.8	-	4.33	9.3	-	0.56	-	-	-	-	-

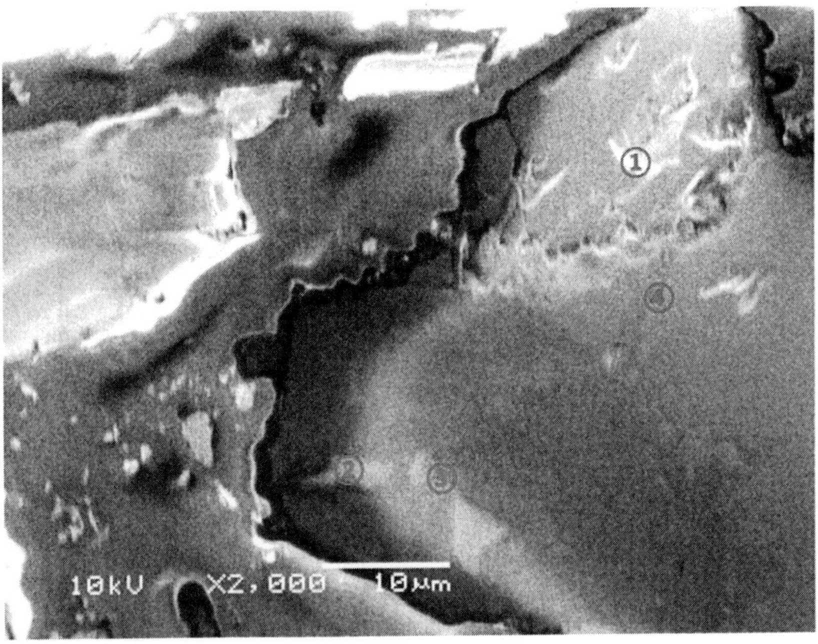

Figure 5. Scanning electron microscopic image of the raw monazite-bearing ore (3000× magnification). The detailed investigations are given in the text and Table 3.

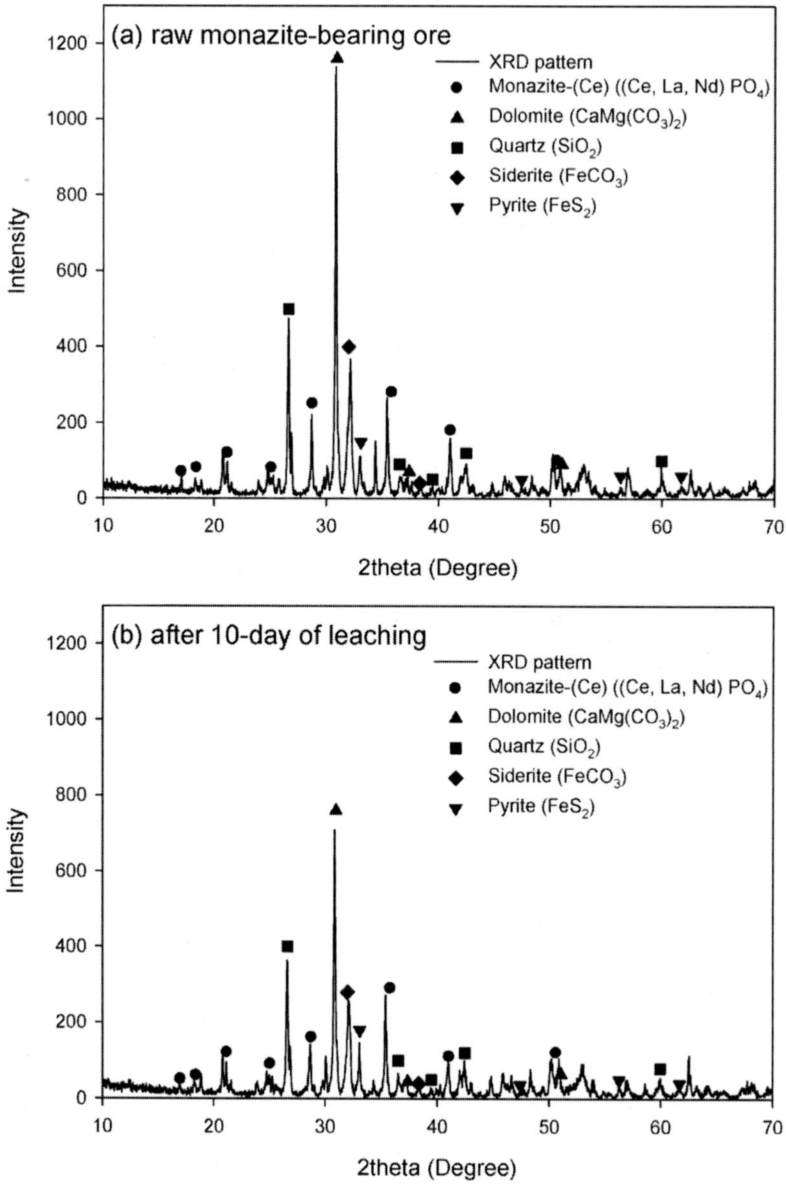

Figure 6. X-Ray diffraction (XRD) patterns of the monazite-bearing ore before and after leaching (10 days).

Biological leaching technology has some advantages over other methods, as it is relatively simple, inexpensive, and environmental friendly. However, even though *A. aceti* showed the highest leaching efficiency of the PSB tested, the efficiency of extraction was only ~0.13% (at least in batch-type experiments), which is still very low. The PSB used in this study have been used in soil improvement applications because they adapt well to soil environments; however, their organic acid producing and phosphate solubilizing ability is not that high. Thus, further studies are underway to explore the use of other microorganisms with good organic acid producing abilities for REE extraction applications, as well as to design continuous leaching systems so as to enhance leaching efficiency because the ore particles might not be exposed or contacted by the organic acid produced by the test PSB strains in a batch culture system.

4. CONCLUSIONS

The present study was conducted to use PSB to leach REE from monazite-bearing ore and to predict the REE leaching ability of PSB by determining halo zone formation on agar media and phosphate solubilization in liquid media. Among the ten test PSB strains, the phosphate solubilizing ability of *A. aceti* was the highest, based on results of halo zone formation and the phosphate solubilization test. Based on REE leaching data from the raw monazite-bearing ore by the test PSB, halo zone formation was found to be a useful method to select high-capacity bacteria in REE leaching.

ACKNOWLEDGMENTS

This research was supported by the Basic Research Project (GP2015-005 and JP2013-005) of the Korea Institute of Geoscience and Mineral Resources (KIGAM), funded by the Ministry of Science, Information and Communications Technologies, and Future Planning of Korea.

AUTHOR CONTRIBUTIONS

All of the authors were involved in the laboratory work, sample characterization, writing and revising of all parts of the manuscript.

CONFLICTS OF INTEREST

The authors declare no conflict of interest.

REFERENCES

1. British Geological Survey. Rare Earth Elements. Available online: http://www.MineralsUK.com (accessed on 27 March 2015).
2. Gupta, C.K.; Krishnamurthy, N. *Extractive Metallurgy of Rare Earths*; CRC Press: New York, NY, USA, 2005.
3. Maestro, P.; Huguenin, D. Industrial applications of rare earths: Which way for the end of the century. *J. Alloy. Compd.* **1995**, *225*, 520–528. [Google Scholar] [CrossRef]
4. Habashi, F. *Handbook of Extractive Metallurgy*; Wiley-VCH: Heidelberg, Germany, 1997.
5. Zongsen, Y.; Minbo, C. *Rare Earth Elements and Their Applications*; Metallurgical Industry Press: Beijing, China, 1995.
6. Park, H.-K.; Lee, J.-Y.; Cho, S.-W.; Kim, J.-S. Overview on the Technologies for Extraction of Rare Earth Metals. *J. Korean Inst. Resour. Recycl.* **2012**, *21*, 74–83.
7. Desouky, O.A.; El-Mougith, A.A.; Hassanien, W.A.; Awadalla, G.S.; Hussien, S.S. Extraction of some strategic elements from thorium–uranium concentrate using bioproducts of *Aspergillus ficuum* and *Pseudomonas aeruginosa*. *Arabian J. Chem.* **2011**.

8. Hassanien, W.A.G.; Desouky, O.A.N.; HussienN, S.S.E. Bioleaching of some Rare Earth Elements from Egyptian Monazite using *Aspergillus ficuum* and *Pseudomonas aeruginosa*. *Walailak J. Sci. Technol.* **2014**, *11*, 809–823.

9. Goyne, K.W.; Brantley, S.L.; Chorover, J. Rare earth element release from phosphate minerals in the presence of organic acids. *Chem. Geol.* **2010**, *278*, 1–14.

10. Chung, H.; Park, M.; Madhaiyan, M.; Seshadri, S.; Song, J.; Cho, H.; Sa, T. Isolation and characterization of phosphate solubilizing bacteria from the rhizosphere of crop plants of Korea. *Soil. Biol. Biochem.* **2005**, *37*, 1970–1974.

11. Gadagi, R.S.; Sa, T. New isolation method for microorganisms solulbilizing iron and aluminum phosphates using dyes. *Soil Sci. Plant Nutr.* **2002**, *48*, 615–618.

12. Gadd, G.M. Fungal Production of Citric and Oxalic Acid: Importance in Metal Speciation, Physiology and Biogeochemical Processes. In *Advances in Microbial Physiology*; Poole, R.K., Ed.; Academic Press: Waltham, MA, USA, 1999; Volume 41, pp. 47–92.

13. Chen, Y.P.; Rekha, P.D.; Arun, A.B.; Shen, F.T.; Lai, W.A.; Young, C.C. Phosphate solubilizing bacteria from subtropical soil and their tricalcium phosphate solubilizing abilities. *Appl. Soil Ecol.* **2006**, *34*, 33–41.

14. Rodriguez, H.; Fraga, R. Phosphate solubilizing bacteria and their role in plant growth promotion. *Biotechnol. Adv.***1999**, *17*, 319–339.

15. Jeong, S.; Moon, H.S.; Nam, K.; Kim, J.Y.; Kim, T.S. Application of phosphate-solubilizing bacteria for enhancing bioavailability and phytoextraction of cadmium (Cd) from polluted soil. *Chemosphere* **2012**, *88*, 204–210.

16. Peix, A.; Rivas, R.; Mateos, P.F.; Martínez-Molina, E.; Rodríguez-Barrueco, C.; Velázquez, E. *Pseudomonas rhizosphaerae*sp. nov., a novel species that actively solubilizes phosphate *in vitro*. *Int. J. Syst. Evol. Microbiol.* **2003**, *53*, 2067–2072.

17. Gholami, A.; Shahsavani, S.; Nezarat, S. The Effect of Plant Growth Promoting Rhizobacteria (PGPR) on Germination, Seedling Growth and Yield of Maize. *Int. J. Biol. Life Sci.* **2009**, *1*, 35–40.

18. Tilak, K.V.B.R.; Ranganayaki, N.; Pal, K.K.; De, R.; Saxena, A.K.; Nautiyal, C.S.; Mittal, S.; Tripathi, A.K.; Johri, B.N. Diversity of plant growth and soil health supporting bacteria. *Curr. Sci.* **2005**, *89*, 136–150.

19. Halder, A.K.; Chakrabartty, P.K. Solubilization of inorganic phosphate by *Rhizobium*. *Folia Microbiol.* **1993**, *38*, 325–330.

20. Reyes, I.; Bernier, L.; Simard, R.R.; Tanguay, P.; Antoun, H. Characteristics of phosphate solubilization by an isolate of a tropical *Penicillium rugulosum* and two UV-induced mutants. *FEMS Microbiol. Ecol.* **1999**, *28*, 291–295.

21. Briggs, P.H.; Meier, A.L. *The Determination of Forty Two Elements in Geological Materials by Inductively Coupled Plasma-Mass Spectrometry*; US Department of the Interior, US Geological Survey: Denver, CO, USA, 1999.

22. Katznelson, H.; Peterson, E.A.; Rouatt, J.W. Phosphate-dissolving microorganisms on seed and in the root zone of plants. *Can. J. Bot.* **1962**, *40*, 1181–1186.

23. Stumm, W.; Morgan, J.J. *Aquatic Chemistry: Chemical Equilibria and Rates in Natural Waters*; John Wiley & Sons: Hoboken, NJ, USA, 2012.

24. Hwangbo, H.; Park, R.D.; Kim, Y.W.; Rim, Y.S.; Park, K.H.; Kim, T.H.; Suh, J.S.; Kim, K.Y. 2-Ketogluconic acid production and phosphate solubilization by *Enterobacter intermedium*. *Curr. Microbiol.* **2003**, *47*, 87–92.

25. Švitel, J.; Šturdik, E. 2-Ketogluconic acid production by *Acetobacter pasteurianus*. *Appl. Biochem. Biotechnol.* **1995**, *53*, 53–63.

26. Dawson, R.M.C. *Data for Biochemical Research*; Clarendon Press: Oxford, UK, 1959.

27. Qin, F.; Shan, X.-Q.; Wei, B. Effects of low-molecular-weight organic acids and residence time on desorption of Cu, Cd, and Pb from soils. *Chemosphere* **2004**, *57*, 253–263.

28. Debela, F.; Arocena, J.M.; Thring, R.W.; Whitcombe, T. Organic acid-induced release of lead from pyromorphite and its relevance to reclamation of Pb-contaminated soils. *Chemosphere* **2010**, *80*, 450–456.

29. Bolan, N.S.; Naidu, R.; Mahimairaja, S.; Baskaran, S. Influence of low-molecular-weight organic acids on the solubilization of phosphates. *Biol. Fertil. Soils* **1994**, *18*, 311–319.

30. Pine, L.; George, J.R.; Reeves, M.W.; Harrell, W.K. Development of a chemically defined liquid medium for growth of*Legionella pneumophila*. *J. Clin. Microbiol.* **1979**, *9*, 615–626.

14

CHAPTER

Porosity and Permeability of Round Top Mountain Rhyolite (Texas, USA) Favor Coarse Crush Size for Rare Earth Element Heap Leach

Lorraine Negron [1], Nicholas Pingitore [1,*] and Daniel Gorski [2]

[1]*Department of Geological Sciences, The University of Texas at El Paso, El Paso, TX 79968-0155, USA*

[2]*Texas Rare Earth Resources, 539 El Paso St., Sierra Blanca, TX 79851, USA*

ABSTRACT

Water-saturation porosity and dye-penetration permeability measurements of Round Top Mountain rhyolite confirm that a ½-inch (13-mm) crush size would permit efficient acid heap leaching of yttrium and heavy rare earth elements (YHREEs) hosted in yttrofluorite, a YHREE-substituted variety of fluorite. Laboratory acid leaching has extracted up to 90% of the YHREEs. The bulk insoluble gangue mineralogy of the rhyolite, 90% to 95% quartz and feldspars, assures low acid consumption. Different crush sizes were weighed, soaked in water, and reweighed over time to determine water-penetration estimated porosity. Typical porosities were 1% to 2% for gray and 3% to 8% for pink varieties of Round Top rhyolite. The same samples were re-tested after soaking in dilute sulfuric to simulate heap leaching effects. Post-leach porosity favorably increased 15% in pink and 50% in gray varieties, due to internal mineral dissolution. Next, drops of water-based writing ink were placed on rhyolite slabs up to ~10 mm thick, and monitored over time for visual dye breakthrough to the lower side. Ink penetration through 0.5 to 2.5-mm-thick slabs was rapid, with breakthrough in minutes to a few hours. Pink rhyolite breakthrough was faster than gray. Thicker slabs, 4 to 10 mm, took hours to three days for breakthrough. Porosity and permeability of the Round Top rhyolite and acid solubility of the yttrofluorite host should permit liberation of YHREEs from the bulk rock by inexpensive heap leaching at a coarse and inexpensive nominal ½-inch (13-mm) crush size. The rate-limiting step in heap leach extraction would be diffusion of acid into, and back-diffusion of dissolution products out of, the crushed particles. The exceptional porosity and permeability that we document at Round Top suggest that there may be other crystalline rock deposits that economically can be exploited by a coarse-crush bulk heap leach approach.

Keywords: heavy rare earth elements; yttrofluorite; heap leach; yttrium; dysprosium; YHREE; rhyolite; Round Top

1. INTRODUCTION

Yttrium and the heavy rare earth elements (YHREEs) are vital components of the materials in the optical, electronic, and mechanical products that define 21st Century technologies [1]. Virtually, all YHREEs currently are sourced from the famed south China ionic clay deposits [2]. Taxation, regulatory policies, and environmental degradation are challenges to the steady supply of those elements. Expanded demand for select YHREEs might exceed supply in the future. Thus, there is a global search for alternative sources beyond China, driven by these factors, as well as by trade policies and economic considerations [3,4].

Round Top Mountain in Hudspeth County, west Texas, USA, is a surface-exposed peraluminous ($Al_2O_3 > Na_2O + K_2O + CaO$) rhyolite laccolith, a mushroom-shaped igneous intrusion. The mountain is roughly 2 km in diameter and over 1200 m high, with a rhyolite mass estimated at some 1.5 billion tonnes (Figure 1) [5]. The rock is enriched in rare earth elements (REEs) in excess of 500 ppm, of which the economically desirable heavy REEs and yttrium

(YHREEs) comprise approximately 72%. The rhyolite intrusion is Late Eocene in age, dated radiometrically at 36 Ma (million years) [6]. Although originally emplaced at shallow depth beneath the Earth's surface, subsequent erosion of the overlying strata has left this rock, more resistant to gradual weathering, as a prominent topographic feature. Thus, there is little or no overburden that would need to be removed to commence surface mining by blasting, moving, crushing, and stacking operations to prepare a heap leach pad.

Figure 1. Round Top Mountain yttrium and the heavy rare earth elements (YHREE) deposit, Hudspeth County, west Texas, USA. Virtually the entire mountain, extending a short distance into the subsurface, is mineralized. Note drill rig on peak for scale.

Round Top is enriched in a suite of incompatible elements, those that are due to their ionic radius and charge are not easily incorporated into common igneous rock-forming minerals. At Round Top, these include YHREEs, light REEs (LREEs), Sn, Be, Li, Cs, Rb, U, Th, Ga, Nb, and Ta. Although a number of REE-bearing minerals have been reported in the deposit, synchrotron-based X-ray absorption spectroscopy documented that yttrofluorite hosts essentially all of the YHREEs [7]. Yttrofluorite is a variety of the common mineral fluorite (CaF_2, isometric) in which the trivalent (3+) Y and HREEs substitute for ~5% to 30% of the bivalent Ca^{2+} cations, yielding the formula $(Ca_{1-x},Y,HREE_x)F_{2+x}$ [8]. No other yttrofluorite-hosted YHREE deposit has yet been described; the

Round Top Mountain deposit mineralogy at present appears to be unique.

Mineralization is pervasive and homogeneous through the laccolith, with the exception of the rhyolite margins, where mineralizing fluids interacted with the enclosing country rock and concentrated or diluted various elements. YHREE mineralization occurred via pervasive deposition from late stage (end of the magma cooling and solidification) fluorine-enriched fluid, in which the incompatible elements had concentrated [5,9,10].

The target YHREE-containing yttrofluorite occurs in the fine-grained matrix or groundmass of the rhyolite. Large crystals (up to perhaps 250 μm in diameter) of feldspars and quartz formed during early slow cooling, and were encased in a fine-grained matrix that formed during rapid cooling when the magma was emplaced near the Earth's surface. Most of the yttrofluorite grains range from perhaps 10 to 1 μm or less in diameter. The small size of the target yttrofluorite grains renders such conventional physical beneficiation techniques as froth flotation, gravity separation, magnetic separation, *etc.*uneconomical, chiefly, due to the extremely fine, and perhaps unachievable, particle size required to liberate the minuscule yttrofluorite crystals by sequential crushing, grinding, and milling. Nonetheless, yttrofluorite is soluble in dilute sulfuric acid and thus the deposit potentially could be exploited by heap leaching [11].

The Round Top deposit is situated on land owned by the State of Texas, currently leased by Texas Rare Earth Resources, Inc. (TRER, Sierra Blanca, TX, USA), a publically traded (US stock ticker TRER) mining exploration and development company. TRER has issued preliminary plans to extract the YHREEs, LREEs, and byproduct U, and possibly Be and Li, via low-cost heap leaching with dilute sulfuric acid [12,13].

In a heap leach operation, mineralized rock is blasted and crushed (and ground or milled to a fine particle size, if required), then placed in a large shallow pit previously lined with an impermeable barrier of clay or synthetic material. Crushed rock is stacked to a height of perhaps 10 m, and then irrigated from above with the solvent (in this case it would be dilute sulfuric acid). The leach solution percolates down through the pile, wetting surfaces of the particles. The solution reacts with both the external surface of the crushed particles, and with the interior of the particles by infiltrating and saturating their contained micropore system. Leach solution is collected from the pit bottom, recharged with acid as needed, and recirculated or drawn off for removal of dissolved target and waste ions in an adjacent chemical processing plant [14].

Bench scale column test leaching of composite samples of Round Top rhyolite with dilute sulfuric acid yielded recoveries of up to 90% of the YHREEs at various crush sizes up to 0.5 inches (~13 mm) [13]. This observation suggests that the acid solvent solution is able to penetrate to the centers of such grains, as well as permeate most of the rock volume of the particle to achieve the nearly complete YHREE recoveries observed in the laboratory. These petrophysical qualities of the rhyolite are thus crucial to scaling the proposed leaching process from bench studies to pilot plant to industrial field operations.

This petrophysical study examines the porosity and micro-permeability of the Round Top rhyolite at a scale, roughly 5 to 10 mm or pebble sized, similar to anticipated heap leach crush sizes. Large voids and fluid paths, such as faults and joints in the massive rock (meter to km scale, boulder to outcrop size), are not germane to the proposed crushing and heap leach operations and thus were not examined. Note that use of the term micro-permeability reflects that the relevant fluid pathways in the crushed material occur chiefly at a micrometer (μm) or smaller scale.

Porosity is defined as the volume of contained void space expressed as a fraction or percent of the total rock volume. This percentage varies dramatically with rock type and individual occurrence; some sedimentary rocks can exhibit 40% or more porosity, whereas igneous rocks, such as granite and rhyolite, typically have porosities in the low single digit range [15,16,17,18,19,20]. For most crystalline rocks (*i.e.*, igneous and metamorphic groups), porosity values are low due to formation of the rock at high temperatures and/or pressures. In some igneous rock commercial deposits, the mineralizing hydrothermal solutions enhanced the original porosity and permeability.

Permeability is the ability of a rock to transmit fluids, typically water, gas, or oil. Unless extensively fractured, most crystalline rocks exhibit low permeabilities [19,21]. Thus, a slab of unbroken granite can be suitable for use as a kitchen counter top but does require a synthetic sealant to prevent minor penetration and staining.

In the petroleum industry, porosity and permeability are conventionally measured by the injection of mercury into a rock specimen under increasing pressure [22]. Although this technique is appropriate for the ambient pressures several or more kilometers deep in the Earth's subsurface, it is not relevant to the mild surface conditions in a heap leach. To simulate anticipated heap leach conditions, we therefore designed experiments based on penetration of water and water-based ink into our rock under Earth-surface temperature and atmospheric pressure. The experiments thus simulate anticipated heap leach conditions (aqueous solutions, ambient temperature and pressure).

2. MATERIALS AND METHODS

2.1. Rhyolite Samples

Two colors, pink and gray, dominate the rhyolite found at Round Top. Previous investigators have described 4 or 5 colors (pink, red, purple, gray, and tan, the latter of limited volume and occurring only in places at the borders of the deposit), with no formal distinctions; here, we use pink to include pink, red, and purple hues (Figure 2) [5,9,10]. Samples of the two colors are similar in elemental composition, with the pink hue resulting from the partial oxidation or rusting of magnetite (Fe_3O_4, isometric) grains [5,9,10,23]. In Round Top Mountain, there are no sharp boundaries between these colors and intermediate shades are common. Because the oxidation may have been controlled by, and/or also altered, porosity and permeability, samples of both colors were tested and compared.

Figure 2. Polished blocks and specimens of pink (including red and purple hues) and gray varieties of Round Top Mountain rhyolite. Note sub-mm size of the mineral grains. Scale = 16 cm.

Samples for analysis were taken from a large (several hundred kg), well-mixed composite of material recovered from numerous reverse circulation bore holes drilled at scattered sites across the deposit [13]. Crush sizes and size ranges were produced by conventional sieve separations or visual inspection and selection of particles on appropriately sized printed measurement grids.

2.2. Porosity Measurements

A total of 15 pink and 15 gray samples were chosen by hand from crushed material based on estimated size. Samples were divided into three particle size categories with 5 samples each: 5 mm × 5 mm (visual grid inspection), 5 to10 mm (sieved), and 10 mm (visual grid). Sample sets, after weighing (between 5 and 11 g), were placed into 50-mL beakers containing a mixture of nine parts de-ionized water and one part tap water (to neutralize pH while limiting dissolved solids), covered, and allowed to soak.

Pre- and post-soak sample weights served to determine the mass of water that penetrated the samples during a 3-week period of immersion. The volume of pore space filled in this process was derived from the added water mass (with the calculation taking into account, of course, the respective densities of water and rhyolite). In order to accurately measure the gain in mass due to water infiltration of the interior pore space of each sample, but not due to water clinging on the exterior of the particle. It was imperative to dry the surface of the sample, while retaining the interior moisture. This was accomplished by quickly blowing air on the rock surface until it was visually dry. Dried samples exhibited a characteristic rough and dusty appearance, in contrast to the shiny clean surface when wet. Inasmuch as some interior moisture could be lost in this process, the weight gain and calculated porosity (more precisely stated, the estimated porosity) thus obtained can be considered a minimum. Nonetheless, repeated measures over time and examination and comparison of 5 multiple samples confirm the accuracy of the technique for determination of estimated porosity. Weights for all samples were recorded after 1, 2, 4, 8, 13, and 21 days.

Next, the Round Top samples were dried for 2 days. Then, samples were placed into beakers with 80 mL of 4.25% (v/v) H_2SO_4 and left on a shaker table for 3 weeks. This simulates the initial, and most productive (the period during which most of the YHREEs are released), stage of the heap leach process, which would actually continue for up to 2 to 3 months. After 3 weeks, samples were rinsed 3 times with water and transferred into a clean beaker. Samples were again left to soak for 24 h in the same type of water used for the porosity measurements to allow the diffusion of residual acid from the interior of the samples. This procedure was repeated twice. Samples were then removed from the water bath and left to air dry for 24 h before recording the new current dry weight. This weight was used to quantify the effects of the acid treatment in a second round of "post-acid-leach" porosity experiments. This procedure to simulate the heap leach would document anticipated increases in porosity due to the corrosive action of the acid. Note, of course, that under the field conditions of an industrial heap leach operation, increases in porosity would occur over considerably longer time scales than those modeled here at bench scale.

2.3. Permeability Measurements

Both pink and gray samples of Round Top rhyolite were cut into parallel-sided slabs at staged thicknesses of up to 10 mm with the use of a low-speed circular saw with a diamond blade (Buehler Isomet™, Lake Bluff, IL, USA) and/or a coarse diamond grinding wheel. A single drop of consumer fountain pen ink (Parker Quink™, Newhaven, UK, washable blue) was placed on each slab and penetration breakthrough to the underside was tracked over time by observing the slab bottom in a mirror placed below the slabs (Figure 3). To prevent the ink drop from spreading laterally, and even bleeding down the sides of the sample slab, prior to placement of the drop, a small barrier ring was created on the slab surface with quick-drying clear fingernail polish. The ink was refreshed as needed, and water added when evaporation was seen to have concentrated and thickened the ink.

Figure 3. Experimental setup to permit simultaneous observation of ink applied to tops of slabs and image in mirror, angled below, of breakthrough ink on the undersides of slabs. Metal gooseneck piece in foreground is a high-intensity lamp.

Some thicker samples (>2.5 mm), particularly of the gray variety, were found to require up to 2 to 3 days for penetration. Therefore, a shallow (<1 mm deep) dimple was drilled into the upper surface to permit placement of a larger drop of ink. The samples were then placed into an airtight box containing water in open beakers that would maintain a high ambient humidity (Figure 4). This would slow evaporation and drying of the ink, which could clog the micropores and invalidate the experiment. Additional ink was added as it was lost to both evaporation and penetration of the rock. Penetration breakthrough and spread were documented photographically and time-recorded for the thin slabs.

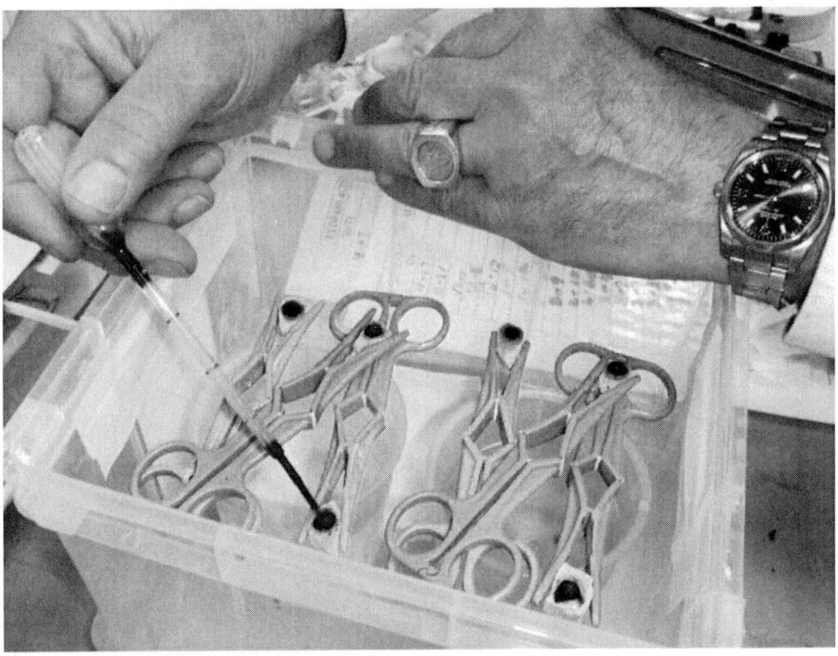

Figure 4. Drops of fountain pen ink being applied to thick slabs of rhyolite in the open humidifying box. Beakers of water to maintain humidity visible at the bottom of box.

3. RESULTS

3.1. Pre-Acid-Leach Porosity

In the first set of trials (pre-acid leach), the pink rhyolites displayed overall higher porosities than the gray rhyolites, for all particle sizes (Figure 5a). Variation among the five individual particles measured in each experimental set was more pronounced for the pink rhyolites than for the grays. These two observations suggest that the magnetite (and possibly other minerals) oxidation reactions that created the pink color also created additional porosity and/or additional micro-permeability that permitted better penetration of the water in our experiments. Similarly, the magnetite oxidation might have been controlled by the prior distribution of permeability and porosity in the nearly solidified rhyolite.

For each of the two individual colors, variation in porosity between particle sizes was relatively minor. We examined different sizes because of the possibility that potential errors in the surface drying process might be correlated with the different surface-area-to-volume ratio inherent in different particle sizes. A further consideration was the possibility that in larger particle sizes (imagine, at the extreme, a boulder) penetration of water to interior porosity might be extremely slow or even impossible, yielding a lower measured water-mass-gain porosity, or that smaller sizes might expose more internal pore space. The data indicate that none of these potential effects was a significant problem in the range of particle sizes tested.

The flattening of the curves of porosity *versus* time (Figure 5a) demonstrates that water saturation of the particles was approached in two to five days. The flattening is more pronounced for the pink rhyolites; some of the gray rhyolites appear to have absorbed minor additional quantities of water throughout the experimental period (slight positive slope to three weeks). The longer saturation time for the gray rhyolite is consistent with their lower porosity.

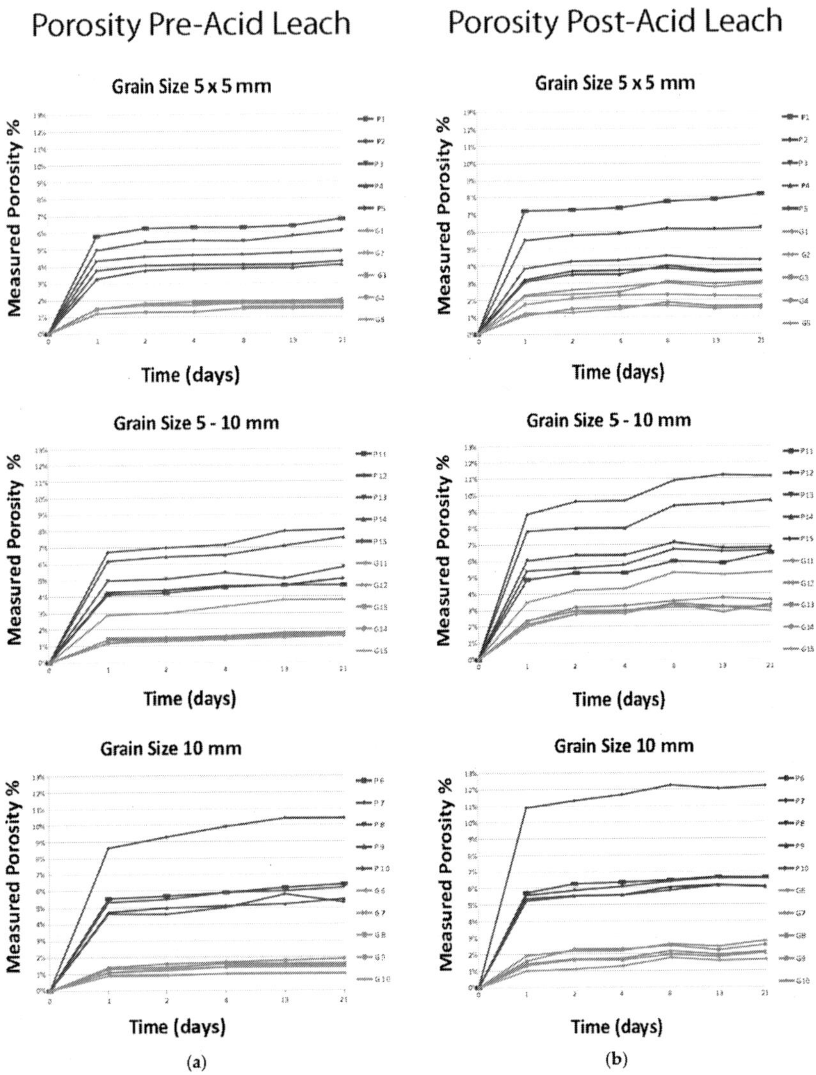

Figure 5. Estimated porosity measurements: (**a**) Pre-acid leach experiments; (**b**) Post-acid leach experiments. Red lines = pink samples; Gray lines = gray samples.

The largest water gain occurred during the first day, an effect confirmed in the permeability studies (Section 3.3.), where mm-scale dye penetration was observed at a time scale of minutes to a few days.

3.2. Post-Acid-Leach Porosity

In the second trial, porosities are seen to have increased overall after leaching with dilute sulfuric acid (Figure 5b). Further, the spread of the data, as seen in the vertical separation of the individual curves, is also greater. This is consistent with the addition of a new variable, the degree of interior corrosion by the acid leach, which potentially increases variation in the data. As in the first set of trials, the pink rhyolites exhibit higher overall porosities than the gray suite.

Interestingly, the approach to water saturation appears consistently to be approximately one week, about twice as long as observed in the pre-acid-leach experiments. This suggests the creation of additional micro-pores by mineral dissolution during the leaching period, which then require additional time to fill.

No porosity trends relative to grain size were evident; the 5 to 10 mm exhibited the highest overall porosities.

The pink rhyolites (all 15 samples used in the study) exhibited an average 15% increase in final porosity (21st day observation) from 6.1% pre-leach to 7.0% post-leach (Figure 6). The gray rhyolites (15 samples) averaged a 50% increase in final porosity, from 1.8% before leaching to 2.7% after leaching.

3.3. Permeability

Penetration of both the pink and gray rhyolites was rapid, and sequential in time with slab thickness (Figure 7). Within 30 min, breakthrough of the ink on all the pink and the gray rhyolite slabs with thickness to 2.5 mm had occurred. Greater thickness, 3 to 10 mm, took up to three days for breakthrough (Figure 8). This is consistent with the two to three days for the similarly sized porosity samples to approach water saturation. Of special significance is the observation that this initial penetration was focused, that is, the initial ink stain penetrations on the slab bottoms were circular and similar in size to those placed on the

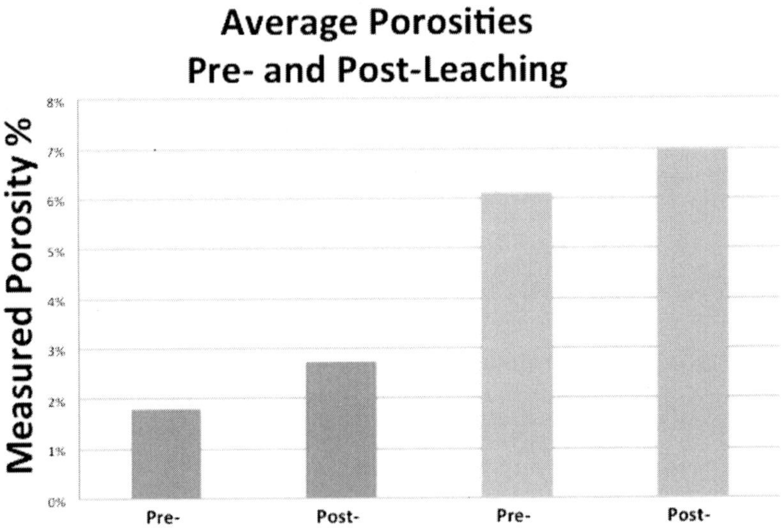

Figure 6. Average (15 samples per color) porosity pre- and post-leaching, showing an increase in porosity after sulfuric acid dissolution. Gray bars represent gray samples; red bars represent pink samples.

slab top. Lateral spreading of the ink at the lower surface of the slabs is only evident at some time well after initial breakthrough. This indicates that the fluid pathways from the top of the slabs to their bottoms were relatively direct. In light of this, it is evident that the micro-pore system in these rhyolites is pervasive and isotropic, with no evidence of preferred orientation of the fluid pathways. On the thick slabs, ink permeation is visible on the sides of many of the blocks, consistent with continued ink spreading during the longer time duration of these experiments.

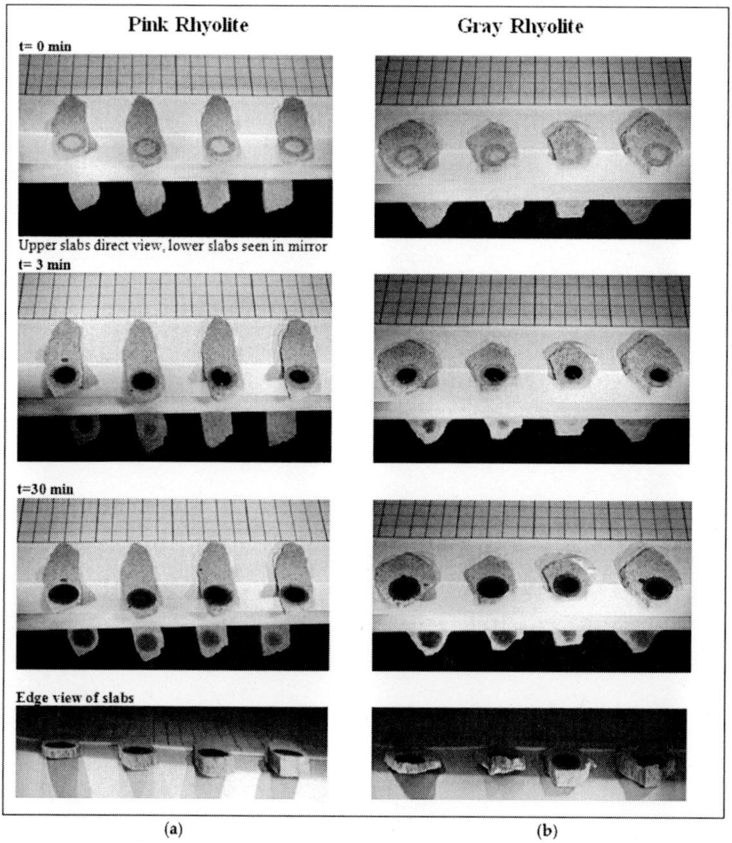

Figure 7. Timed ink penetration of rhyolite slabs to determine permeability. Boxes = 0.5 cm. (**a**) Pink rhyolite; (**b**) Gray rhyolite. Circles seen at $t = 0$ are the ridges created by nail polish, which serve to contain and restrain the ink.

One cautionary note: Once the ink penetrates to the bottom of the slab, it gradually spreads outwards from the transporting pore systems, migrating along the open lower surface. Thus, the ink-stained surface gives the appearance of a denser and more pervasive saturation than actually exists in the interior of the slab. Nonetheless, the stain indicates the general area reached by the micro-pore system in the rhyolite slab.

(a) (b)

Figure 8. (**a**) Top surface of thick slabs with ink dimples; (**b**) Bottom surface of slabs showing ink breakthrough. Exposure times from six hours to three days. Top row of samples, left to right: pink, gray pink; Second row: gray, gray pink; Third row: gray, pink, gray, gray; Bottom row: gray, pink, gray, gray.

In general, the pink variety of rhyolites exhibited more favorable permeability than the gray rhyolite. This was particularly noticeable in tests of the thicker slabs (5 to 10 mm). Pink specimens typically required 6 to 24 h for breakthrough, whereas the grays required as long as three days for penetration.

4. DISCUSSION

4.1. Significance of Round Top Rhyolite Porosity and Permeability

The porosities measured on the Round Top rhyolite samples exceed those of such related rocks as granite, typically reported at 1% or less. Equally important, all samples proved visibly permeable, with rapid penetration and saturation of the particles. These properties contribute to the feasibility of heap leaching at a nominal crush size of ½ inch (13 mm). If the rock were not porous and permeable, it would be necessary to grind or mill to a much finer particle size for the acid to access the minuscule, scattered yttrofluorite target grains. Such mechanical processing would involve considerable expense, in capital equipment, operating costs, and complexity of design. Further, finer particle sizes can lead to jamming of the fluid pathways of the percolating leach fluid, resulting in incomplete exposure of some particles and lower YHREE recoveries.

The ink penetration permeability study demonstrated rapid particle penetration on samples up to 10 mm thick. This suggests that a crush particle 20 mm thick would be efficiently saturated by entry of the solution from all sides of the particle (10 mm radius). Thus, a ½ inch crush size is well within the range of successful testing in this study. It is important to realize that the loci of the YHREEs within the rhyolite apparently are closely associated with the primary porosity of the rock. The minerals hosting the YHREEs and associated incompatible elements were precipitated from a fluorine-rich vapor phase in the final stage of cooling of the magma. These last-stage minerals formed in the space that the vapor phase occupied, and in the pore space opened as the F-vapor corroded pre-existing feldspar crystals [10]. Consequent to this quirk of the rhyolite's genesis, a leach solution entering the micropores of a crush particle will contact most of the target yttrofluorite grains. Very few yttrofluorite grains can be expected to be "locked" in a matrix of acid-insoluble minerals, e.g., quartz and feldspar. This apparent geometric association of porosity and target yttrofluorite contributes significantly to the exceptional recoveries, up to 90%, encountered in both bottle roll and column test heap leach experiments at bench scale [13].

4.2. Heap Leach Efficiency: Pink vs. Gray Rhyolite Varieties

The better performance characteristics of the pink rhyolite in porosity and permeability tests suggests that a heap leach with material solely of this color could achieve faster and perhaps more thorough YHREE recoveries. The distribution of these two varieties in Round Top Mountain would permit selective mining of either color variety. Overall, pink dominates the upper portion of the mountain, gray the center, and pink the lower part. To extract primarily pink, one could mine the deposit from the top, or mine the lower slopes from the level of the surrounding desert. Given the huge size of the mountain, mining solely the pink could be carried out for many decades—tens and tens of years.

Alternatively, the rate of leaching in the heap pile could be slowed or moderated by the judicious mixing of pink and gray rhyolite. Such an approach could be an option if the leach is designed with separate piles, where the leach solution is shunted between different sections to regulate the concentration of ions delivered to the chemical recovery plant. Continuous delivery of a final leach solution of constant composition to the chemical plant is essential to its efficient operation.

4.3. Ionic Diffusion: Rate-Limiting Step for Round Top Heap Leach

The porosity and permeability experiments described herein are consistent with laboratory-scale sulfuric acid leaching experiments described elsewhere, and with unpublished data [13]. Initial dissolution of rare earths and other elements is rapid, with YHREE recoveries of ~33%, in the first 24 h, depending on acid strength and particle size. In bottle roll tests recoveries of 50% to 75% were achieved with 4-mm-diameter particles in one week, and 75% with 10-mm-diameter particles in two weeks. Nonetheless, REE recoveries were observed to increase, albeit with diminishing returns, with continued immersion in acid solution over the course of several to many weeks. Final recoveries of up to 90% for the target YHREEs were achieved in approximately 10 weeks in both bottle roll and column tests.

The time lapse between initial dissolution of YHREEs and near-complete dissolution indicates that transport of acid to the interiors of the particles, and back-transport of the YHREEs out of the particles, is an important rate-limiting step in the leaching process. On the one hand, the permeability breakthrough data and the porosity saturation data indicate that aqueous solutions pass into the rhyolite rapidly, typically taking less than three days to penetrate to the center of a 13-mm-diameter particle (the anticipated crush size for the heap leach). On the other hand, it is obvious that efficient two-way advective transport of the leach solution into and out of the grains is not possible due to the micron to sub-micron dimension of the cracks through which such transport apparently occurs. There is no drive mechanism to push the leach solution in and then back out of these micro-cracks. At crack widths of this scale, capillary force dominates gravity force, and flow through or in and out of the particles does not proceed at economically favorable time scales. Diffusion of acid into the pores and back-diffusion of YHREE and other dissolved species out of the pores will be the most important drive mechanism for YHREE extraction in the heap leach. It is anticipated that diffusion thus will be the rate limiting step in the process. The dynamics of two-way diffusion between aqueous solutions in cracks in rock particle interiors and an external buffering solution has been described in detail in the literature [24].

4.4. Unconventional Opportunity to Heap Leach Coarse-Crushed Crystalline Rock

Bulk heap leaching of relatively unaltered crystalline rock is not a conventional mineral resource extraction technique. Typically, only carefully selected high-grade areas of hard rock mineral deposits are ever mined, crushed and heap leached. Bulk crystalline deposits occasionally can be crushed to a fine particle size and heap leached, e.g., the 5-mm crush size at the Etango uranium project [25]. The exceptional porosity and permeability for a crystalline rock that we documented at Round Top present the opportunity to heap leach the bulk deposit at a coarse crush size. This finding suggests that there may be other crystalline rock deposits that economically can be exploited for rare earths and other elements by a similar strategy.

5. CONCLUSIONS

(1) (1)Water-saturation porosity of Round Top rhyolite ranged from 1%–10%, with the pink variety consistently higher than the gray variety.

(2) (2)Water penetration and saturation was rapid, with most occurring within one day, and nearly complete in two to three days.

(3) (3)Porosity increased 15% (pinks) to 50% (grays) after a three-week dilute sulfuric acid leach.

(4) (4)Ink-penetration permeability was rapid (minutes to hours) for slabs up to 2.5 mm. Thicker slabs, to 10 mm, took as much as several days. Pink rhyolite was consistently more permeable than gray rhyolite.

(5) The rate-limiting step for YHREE recovery in the anticipated heap leach is expected to be diffusion of fresh acid from the dripping heap leach solution into the solution-saturated micropores, and back-diffusion of dissolved elements out of the pores.

In light of these findings, the ½-inch (13 mm) crush size proposed by Texas Rare Earth Resources, Inc. for their heap leach design appears sound, and even conservative. Further, the conclusions suggest that bulk coarse-crush heap leaching of other crystalline rock deposits with favorable porosity and permeability may be economical.

The ultimate success of the proposed Round Top Mountain YHREE heap leach will depend not just on the favorable estimated porosity and permeability documented here. Such other factors as wetting of the particles, clay formation or release, presence of fines, chemical reactivity leading to precipitation of secondary compounds, *etc.* all will play a role [26,27,28,29]. Typical heap leach recoveries for such ore deposits as copper porphyries are in the 60% range, significantly less than the 90% achieved in laboratory scale column tests of Round Top rhyolite. This suggests that tall column field scale tests are a logical next step in evaluating and "de-risking" heap leaching of the Round Top deposit.

ACKNOWLEDGMENTS

The authors thank Texas Rare Earth Resources, Inc. for access to samples and technical data. This project was supported by research contracts 26-8211-12 and 26-8211-16 between Texas Rare Earth Resources, Inc. and the University of Texas at El Paso. Funds for covering the costs to publish in open access were obtained from this source.

AUTHOR CONTRIBUTIONS

Nicholas Pingitore conceived and designed the experiments; Lorraine Negron and Nicholas Pingitore performed the experiments; Nicholas Pingitore and Lorraine Negron analyzed the data; Daniel Gorski contributed reagents/materials/analysis tools; Nicholas Pingitore and Lorraine Negron wrote the paper.

CONFLICTS OF INTEREST

Lorraine Negron declares no conflict of interest. She has not received any compensation from the contracts that supported this research. Nicholas Pingitore serves on the Board of Directors of Texas Rare Earth Resources (TRER). He is not and has never been an employee of TRER, nor has he received any compensation from the research contracts that supported this research. Daniel Gorski is the CEO of TRER and serves on its Board of Directors. The funding sponsors had no role in the design of the study; in the collection, analyses, or interpretation of data; in the writing of the manuscript, and in the decision to publish the results.

ABBREVIATIONS

The following abbreviations are used in this manuscript:

YHREEs: yttrium and heavy rare earth elements;

REEs: rare earth elements;

LREEs: light rare earth elements;

HREEs: heavy rare earth elements;

TRER: Texas Rare Earth Resources.

REFERENCES

1. Chakmouradian, A.R.; Wall, F. Rare earth elements: Minerals, mines, magnets (and more). *Elements* 2012, *8*, 333–340.
2. Peishan, Z.; Zhuming, Y.; Kejie, T.; Xueming, Y. *Mineralogy and Geology of Rare Earths in China*; Science Press: Beijing, China, 1995; p. 209.
3. Mariano, A.N. Economic geology of rare earth elements. In *Geochemistry and Mineralogy of Rare Earth Elements. Reviews in Mineralogy 21*; Lipin, B.R., McKay, G.A., Eds.; Mineralogical Society of America: Washington, DC, USA, 1989; pp. 309–337.
4. Mariano, A.N.; Mariano, A., Jr. Rare earth mining and exploration in North America. *Elements* **2012**, *8*, 369–376.
5. Price, J.G.; Rubin, J.N.; Henry, C.D.; Pinkston, T.L.; Tweedy, S.W.; Koppenaal, D.W. Rare-metal enriched peraluminous rhyolites in a continental arc, Sierra Blanca Area, Trans-Pecos Texas; chemical modification by vapor-phase crystallization. In *Ore Bearing Granite Systems; Petrogenesis and Mineralizing Processes (Special Paper (Geological Society of America))*; Stein, H.J., Hannah, J.L., Eds.; Geological Society of America: Boulder, CO, USA, 1990; pp. 103–120.
6. Henry, C.D.; McDowell, F.W.; Price, J.G.; Smyth, R.C. *Compilation of Potassium-Argon Ages of Tertiary Igneous Rocks, Trans-Pecos, Texas*; University of Texas, Bureau of Economic Geology: Austin, TX, USA, 1986; pp. 1–34.
7. Pingitore, N., Jr.; Clague, J.; Gorski, D. Round Top Mountain (Texas, USA), a massive, unique Y-bearing-fluorite-hosted heavy rare earth element (HREE) deposit. *J. Rare Earths* **2012**, *32*, 90–96.
8. Yttrofluorite, Mindat.org. Available online: http://www.mindat.org/gallery.php?min=4371 (accessed on 27 December 2015).

9. Rubin, J.N.; Price, J.G.; Henry, C.D.; Koppenaal, D.W. Cryolite-bearing and rare metal-enriched rhyolite, Sierra Blanca Peaks, Hudspeth County, Texas. *Am. Mineral.* **1987**, *72*, 1122–1130.

10. O'Neill, L.C. REE-Be-U-F Mineralization of the Round Top Laccolith, Sierra Blanca Peaks, Trans-Pecos, Texas. Master's Thesis, University of Texas, Austin, TX, USA, 2014.

11. Crawford, J. Solubility Data on 646 Common and Not so Common Minerals, 2009. Available online: http://www.mindat.org/article.php/553/Solubility+Data+on+646+Common+and+Not+So+Common+Minerals (accessed on 27 December 2015).

12. Texas Rare Earth Resources. Available online: http://www.trer.com (accessed on 27 December 2015).

13. Gustavson Associates. NI 43–101 Preliminary Economic Assessment, Round Top Project, Sierra Blanca, Texas. 2012. Available online: http://trer.com/wp-content/uploads/2012/06/TRER_NI%2043-101%20PEA_KLG_037.pdf (accessed on 27 December 2015).

14. Dhawan, J.; Safarzadeh, M.S.; Miller, J.D.; Rajamani, R.K.; Moats, M.S. Insights into heap leach technology. In Proceedings of the 2012 SME Annual Meeting and Exhibit 2012, Washington, DC, USA, 19–22 February 2012; pp. 560–567.

15. Akinyemi, O.A.; Alabi, A.A.; Ojo, A.I.; Adewusi, O.E. Characterization of density and porosity of rocks samples from Ogun State of Nigeria. *Earth Sci. Res.* **2012**, *1*, 137–146.

16. Bongiolo, E.M.; Bongiolo, D.E.; Sardini, P.; Mexias, A.S.; Siitari-Kauppi, M.; Gomes, M.E.B.; Formoso, M.L. Quantification of porosity evolution from unaltered to propylitic-altered granites: The 14C-PMMA method applied on the hydrothermal system of Lavras do Sul, Brazil. *Ann. Acad. Bras. Ciênc.* **2007**, *79*, 503–517.

17. Takarli, M.; Prince-Agbodjan, W. Temperature effects on physical properties and mechanical behavior of granite:

Experimental investigation of material damage. *J. ASTM Int.* **2008**, *5*, 1–13.

18. Petford, N. Controls on primary porosity and permeability development in igneous rocks. In *Hydrocarbons in Crystalline Rocks*; Petford, N., McCaffrey, K.J.W., Eds.; Geological Society Publishing House: Bath, UK, 2003; pp. 93–107.

19. Brace, W.F. Permeability of crystalline rocks: New *in situ* measurements. *J. Geophys. Res.* **1984**, *89*, 4327–4330.

20. Ilankoon, I.; Neethling, S. The effect of particle porosity on liquid holdup in heap leaching. *Miner. Eng.* **2013**, *45*, 73–80.

21. Sausse, J.; Genter, A. Types of permeable fractures in granite. In *Petrophysical Properties of Crystalline Rocks*; Harvey, P.K., Brewer, T.S., Pezard, P.A., Petrov, V.A., Eds.; Geological Society Publishing House: Bath, UK, 2005; pp. 1–14.

22. Monicard, R.P. *Properties of Reservoir Rocks: Core Analysis*; Éditions Technip: Paris, France, 1980; pp. 1–168.

23. Pingitore, N.; Clague, J.; Gorski, D. Uniform distribution of yttrium and heavy rare earth elements in Round Top Mountain rhyolite deposit, Sierra Blanca Texas, USA: Data, significance, and origin. In Proceedings of the American Geophysical Union, Fall Meeting, San Francisco, CA, USA, 15–19 December 2014.

24. Pingitore, N. The role of diffusion during carbonate diagenesis. *J. Sed. Petrol.* **1982**, *52*, 27–29.

25. Bannerman Resources. Available online: http://32uo991mljzg3wrz8dkq0et1.wpengine.netdna-cdn.com/wp-content/uploads/2015/07/Etango-Heap-Leach-Demonstration-Program-Phase-1-Report.pdf (accessed on 12 December 2015).

26. Bartlett, R. *Solution Mining: Leaching and Fluid Recovery of Materials*, 2nd ed.; Gordon and Breach: Amsterdam, The Netherlands, 1998; p. 470.

27. De Andrade Lima, L. Liquid axial dispersion and holdup in column leaching. *Miner. Eng.* **2006**, *19*, 37–47.

28. Petersen, J.; Dixon, D. Modelling zinc heap bioleaching. *Hydrometallurgy* **2007**, *85*, 127–143.

29. Ilankoon, I.; Neethling, S. Hysteresis in unsaturated flow in packed beds and heaps. *Miner. Eng.* **2012**, *35*, 1–8.

Index

A

Albitite-Greisen, 195

B

bioleaching, 256, 301

C

Ca removal, 212
Cerium Redox, 128, 152
Chromatography, 96, 103, 109, 110, 111
critical metals, 36, 37, 39, 45, 49

D

dissolution behavior, 177, 178, 179, 183, 187, 190
dysprosium, 3, 62, 67, 106, 180, 183, 278

E

East Kazakhstan, 195, 206, 207, 208
elements, 2, 3, 5, 6, 7, 8, 9, 11, 293, 294, 295, 296, 297, 298, 300
Environment, 61, 72, 91, 92, 93, 125, 155, 160, 161, 191, 193

F

Fertilizers, 61, 67
fluorosurfactant, 239, 240, 249

G

Greece, 35, 36, 37, 38, 39, 40, 58, 59, 64, 73
Ground Water, 78

H

HDEHP, 95, 96, 101, 104
homogeneous, 239, 240, 241, 242, 248, 249, 251, 252, 253, 280
hydrometallurgy, 237, 240

K

Kaolinite, 114, 121, 164, 168, 169, 174

L

life cycle assessment, 2, 29

M

Magnetic Separation, 114, 125, 176, 218
Metasomatits, 195
mineralogy, 2, 4, 5, 10, 12, 39, 49, 55, 56, 142, 213, 277, 280
Model Calibration, 96, 105
Multi-Objective, 96, 111

O

Optimization, 96, 100, 104, 106, 111, 211

P

Pegmatite, 195, 199, 209
phosphate, 12, 49, 50, 67, 83, 266, 267, 269, 272, 274, 275

R

rare earth, 2, 3, 4, 6, 7, 8, 9, 10, 257, 277, 278, 279, 297, 298, 300
recycling, 2, 3, 4, 6, 18, 19, 33, 36, 53, 56, 70, 178, 239, 240, 242, 248, 249
Rees, 61
Reserves, 61, 64
Reutilize, 164

S

Seawater, 72, 128, 133, 136, 143, 152, 153, 154, 155, 156, 157
Sediment, 128, 139, 142, 153, 155, 159, 208
Separation, 11, 109, 110, 111, 250, 253, 259
sustainable development, 35, 36, 39, 52

T

Tailings, 114, 116, 121, 165, 167, 169

U

unreacted core model, 177, 178, 187, 188, 190

W

West Bank, 77, 78, 79, 80, 81, 92
WHO Limits, 78

Y

yttrofluorite, 277, 278, 279, 280, 292, 293